P9-APJ-694

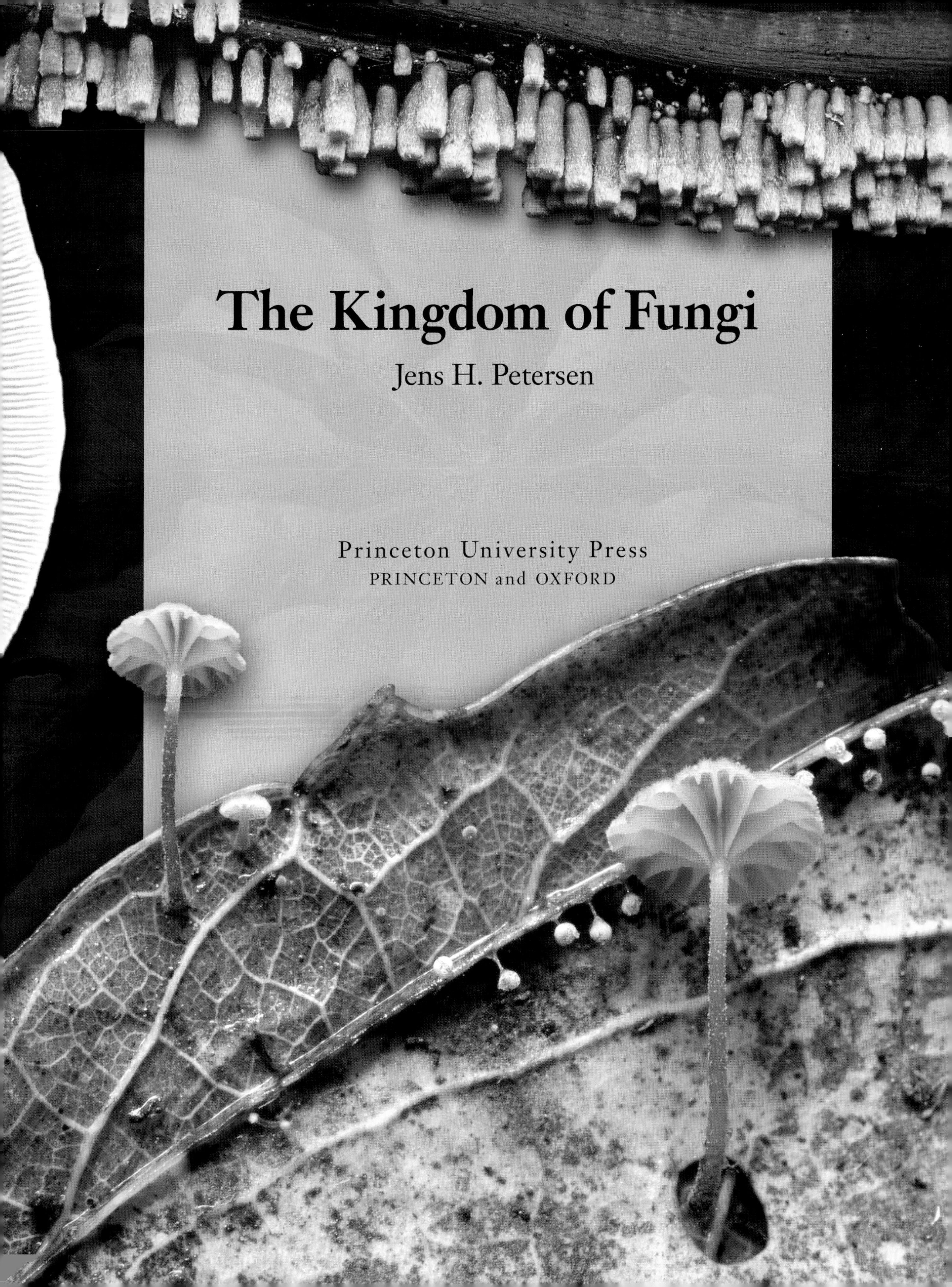

The Kingdom of Fungi

Jens H. Petersen

Princeton University Press
PRINCETON and OXFORD

Contents

Preface

Fungi are everywhere: in forests and fields, in soil, in our buildings, and in the biotech corporations who try to transform straw into biofuel.

Fungi are a source of food and fascination to people around the world, but they are also a mystery, living a hidden life, appearing and disappearing in strange and unpredictable ways.

The kingdom of fungi has been called "the hidden kingdom." It is the last great unknown among the multicellular organisms.

With this book I will try to reveal this kingdom and its fascinating inhabitants.

Introducing fungal life

A super short trip through a fungal generation could look like this:

When a fungal spore germinates it will form long, cylindrical, branched cells called *hyphae*. These grow in a nutritive substrate (soil, wood, dung, etc.), where they form a *mycelium* (below).

The hyphae exude enzymes that disperse into the substrate, where organic material is decomposed. The resulting smaller molecules (e.g., sugars) then diffuse back inside the hyphae and serve as fuel for the growth.

While the fungus grows, it accumulates energy in the mycelium and spreads to new food sources, often by means of thicker *hyphal strings* (facing page, below). After a period of time (weeks, months, years), the mycelium may start to produce *fruiting bodies*.

Fungal fruiting bodies can have all kinds of sizes and shapes, from tiny flask- and cup-shaped structures to the large, well-known agarics, puffballs, polypores, and coral fungi (facing page, top). The purpose of fruiting bodies is to enable the sexual blending of two or more genetic types and produce *spores* for dispersal.

The spores are normally dispersed by wind and once a spore has landed on a suitable substrate, it germinates and produces a new fungal mycelium. Obviously, the majority of spores never get that far.

spores

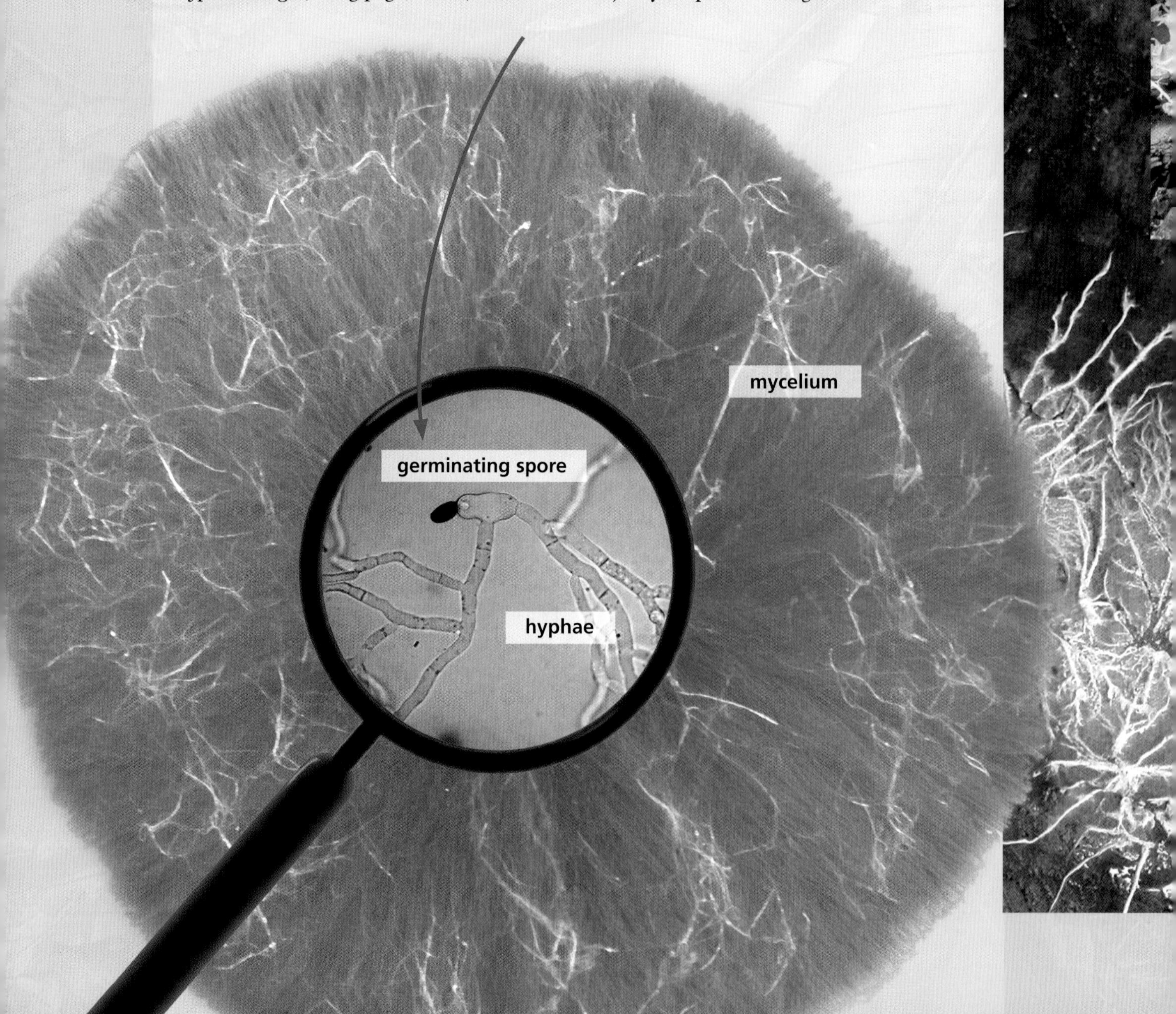

fruiting bodies
hyphal strings

Fungal spores

A spore is a small structure that is fundamental to fungal dispersal. It typically consists of one or a few cells and doesn't contain specialized cells with nutrients, as is the case in plant seeds. Compared to seeds, fungal spores are very, very small, typically about 1/100 of a millimeter long (usually expressed as 10 µm). They may even be smaller—for example, 3 µm—and are extremely well suited for wind dispersal.

There is an amazing variation in spore morphology: from colorless to black, smooth, warty, striated, spiny, crested, and with every imaginable shape. Some reasons for this variation are obvious: spores dispersed in water often have appendices to enable them to float better, while spores destined to germinate on dung are often thick-walled so they can pass through animals' digestive tracts unharmed. But for most spore variation, we have no good explanation—spores just vary beautifully.

Ciboria betulae releasing spores, Denmark (× 15)

Leucocoprinus cretaceus releasing spores, Borneo (TL)

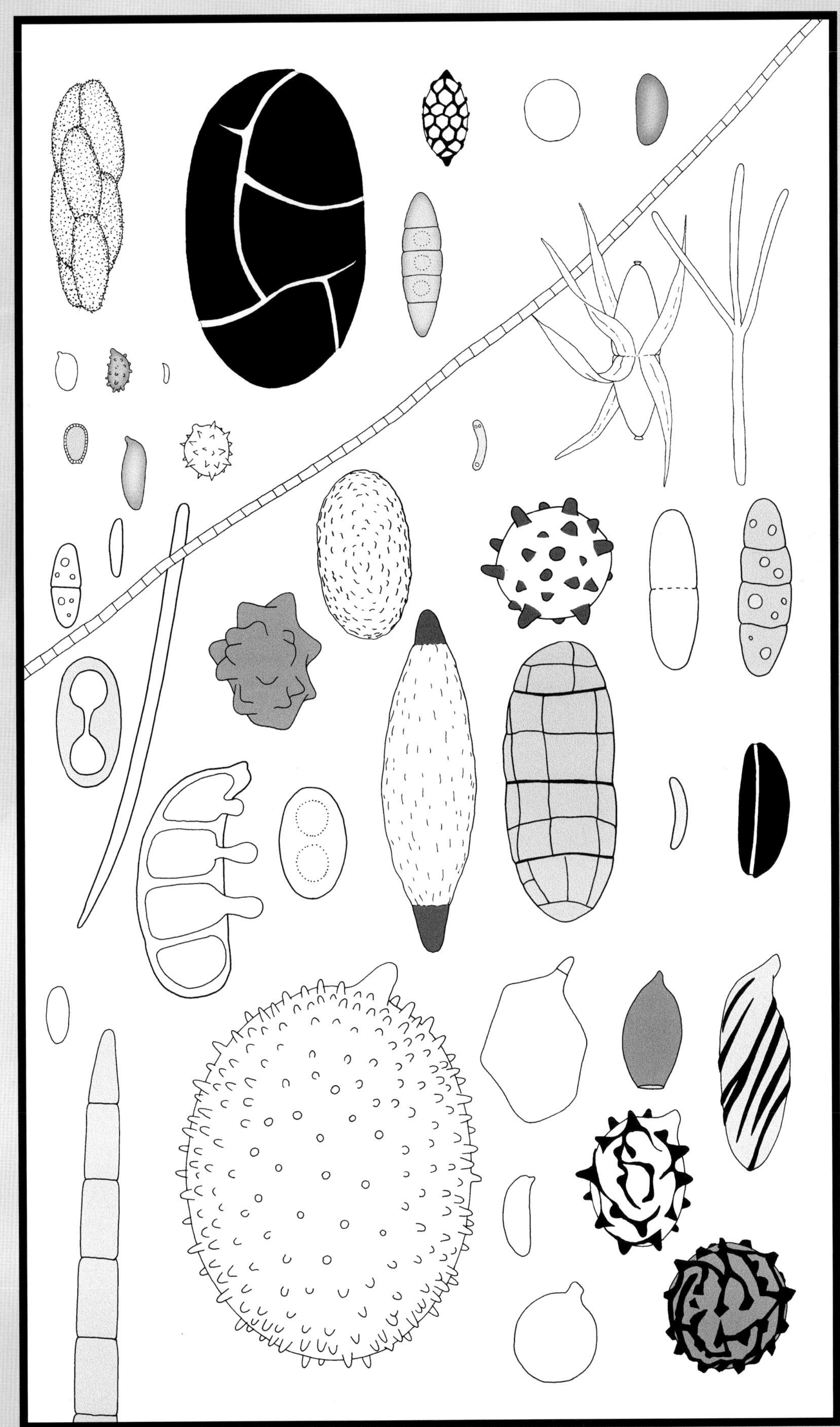

Fungal spore variation (AH)

The ascomycote *Ascobolus sacchariferus* produces spores inside *asci* at the top of the fruiting body. The mature asci protrude and resemble glass cylinders with eight spores floating inside.

When ripe, the asci open at the top and the spores are discharged with great force. In the case of the picture on the facing page, they hit a plastic lid covering the box holding the fungus.

Ascobolus sacchariferus, Denmark (× 50)

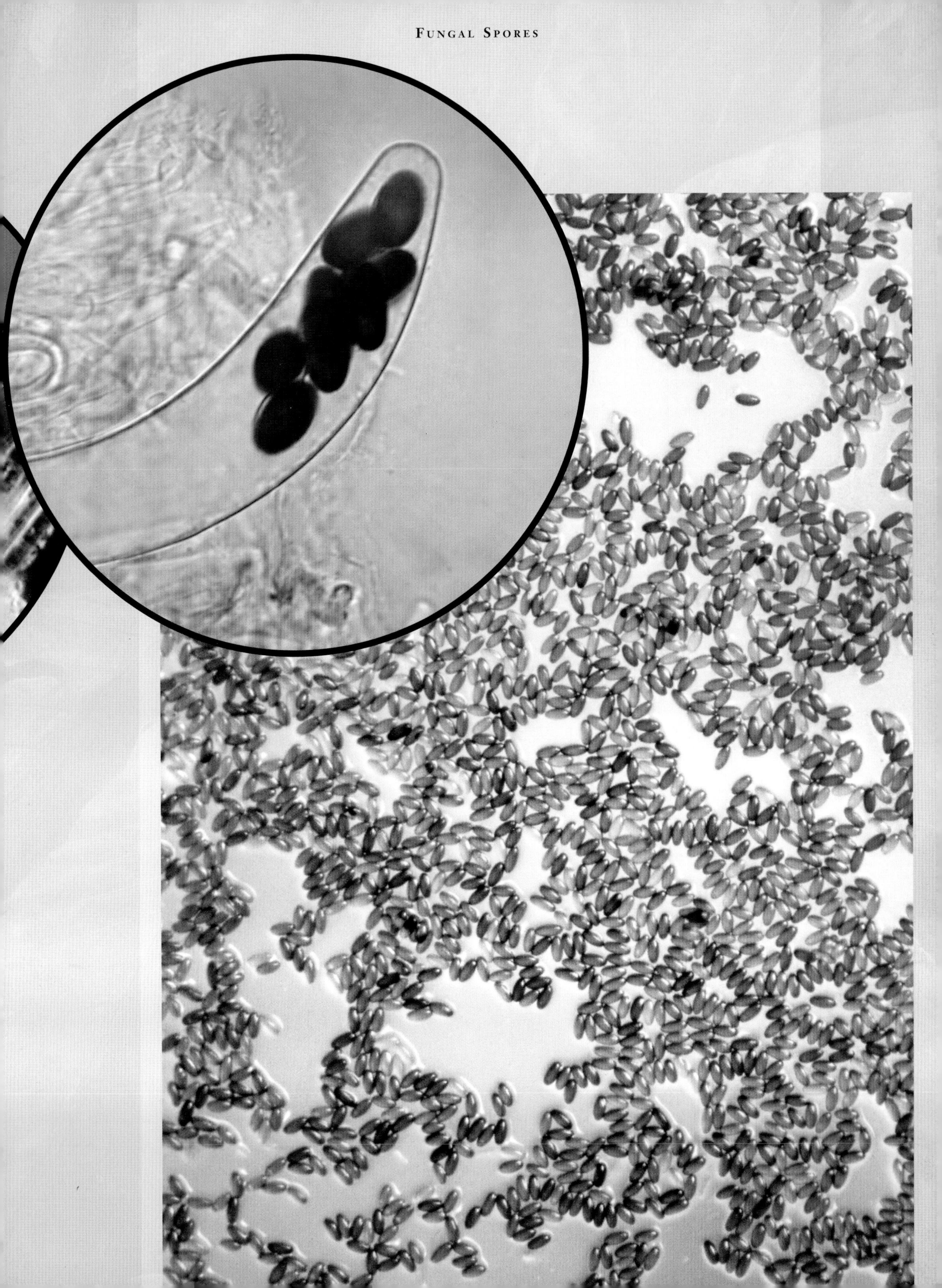

Hyphae

Most fungal cells are much longer than wide and are fused to each other only at the ends. The resulting cell strands are called *hyphae*, and the separations that divide them are called *septa*. Most hyphae are very narrow—5–15 µm—much thinner that a typical plant cell.

Hyphae will branch and form a complex network of cells called a *mycelium*. The picture below shows hyphae germinating from a group of dark spores on a petri dish. As the hyphae spread over the agar, the hyphal tips exude enzymes that decompose nutrients into simple molecules, which can then diffuse into the hyphae, serving as fuel for further growth (see also page 194).

Later in the life cycle of the fungus it may produce fruiting bodies. Although these seem quite robust, they are built from interwoven hyphae—almost like a ball of yarn.

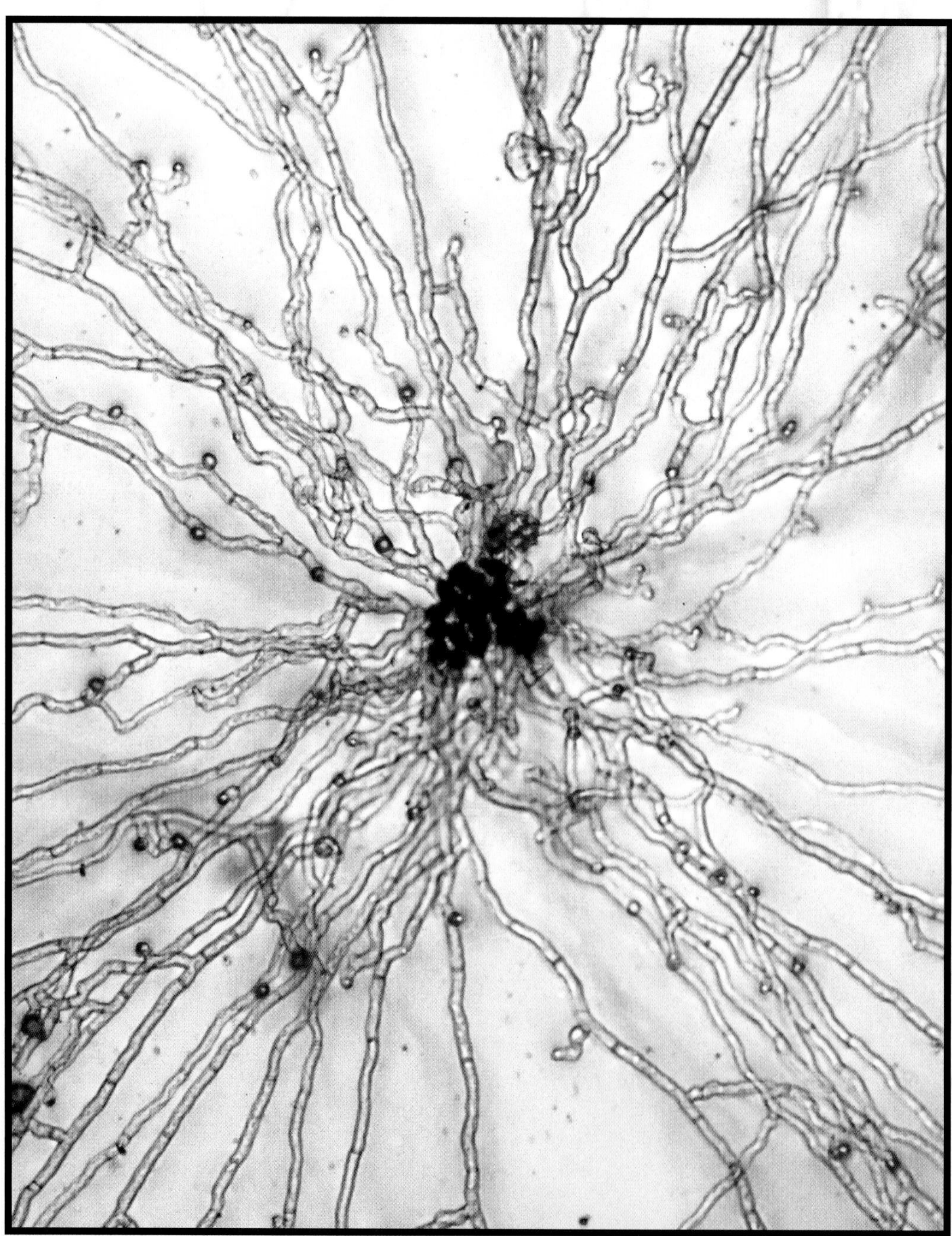

An agar plate with black spores germinating into hyphae (× 250)

To easily move around in their habitat, fungi may form hyphal strings. These are bundles of hyphae in which the outer layers protect the inner from desiccation and stress.

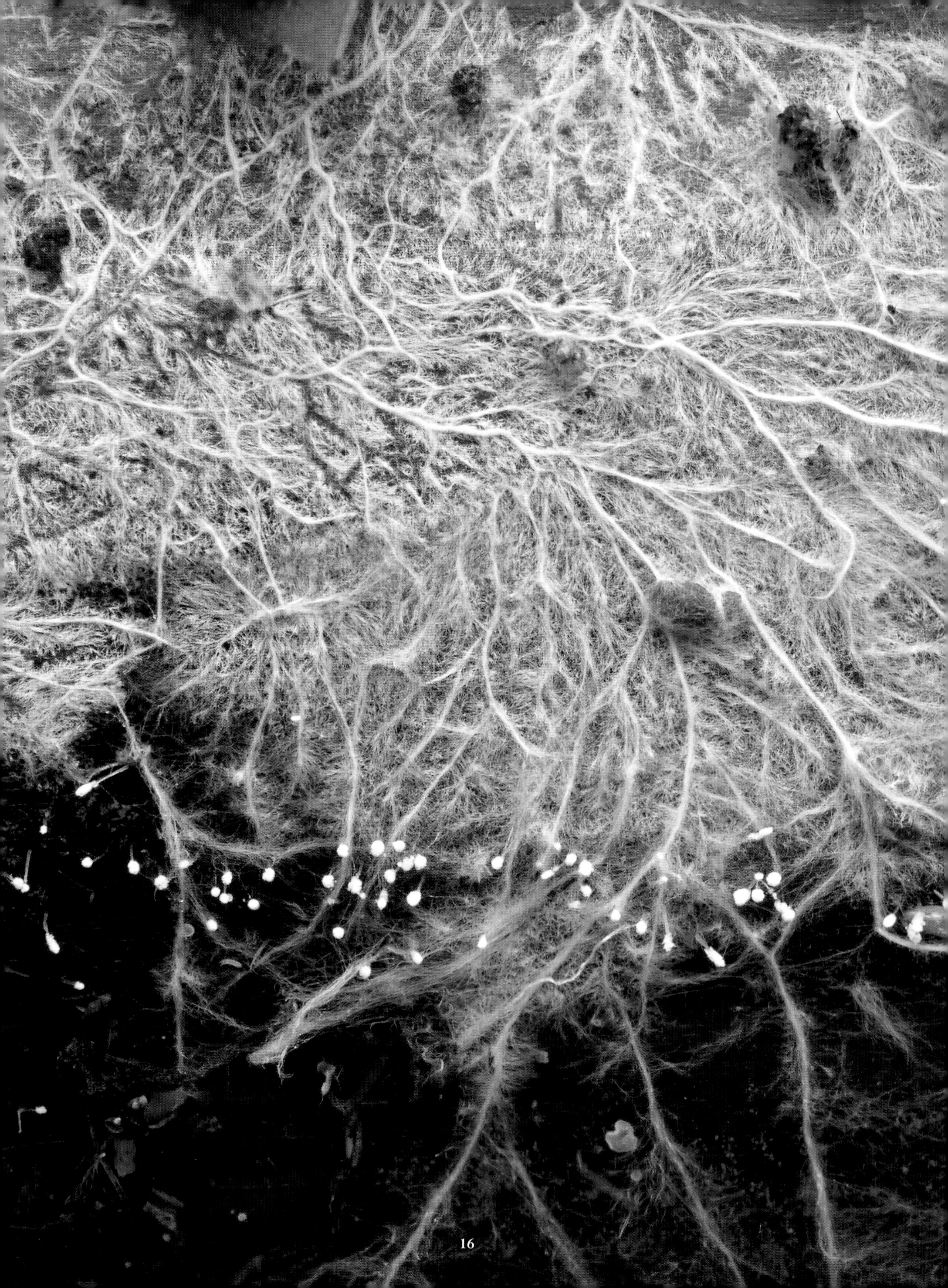

A strong mycelium with hyphae and hyphal strands overtaking other less competitive fungi.

Kinship

Biologists have invented a hierarchical system that places all organisms into named groups according to their kinship. As an overview of this complex system, these *kinship groups* (also called *phylogenetic groups*) are listed below:

The lowest level, *the species*, is—at least in theory—a real entity consisting of individuals that can interbreed. All higher levels are abstractions created to articulate the assumed level of kinship.

domain

kingdom

phylum

class

order

family

genus

species

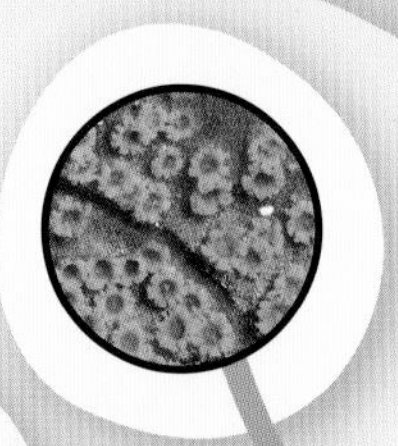

Form groups and kinship groups

In practical work with organisms we often define groups based on their morphology rather than kinship. These highly practical groups are called *form groups* and should not be confused with kinship groups. The red line below, for example, delimits a form group called gasteroid fungi: Basidiomycota that produce spores inside and release them passively.

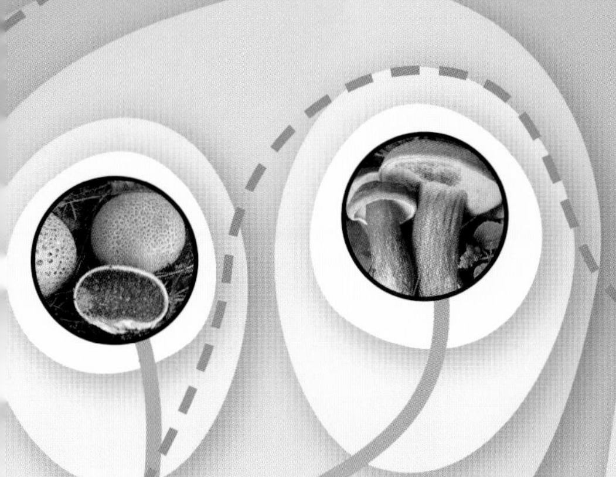
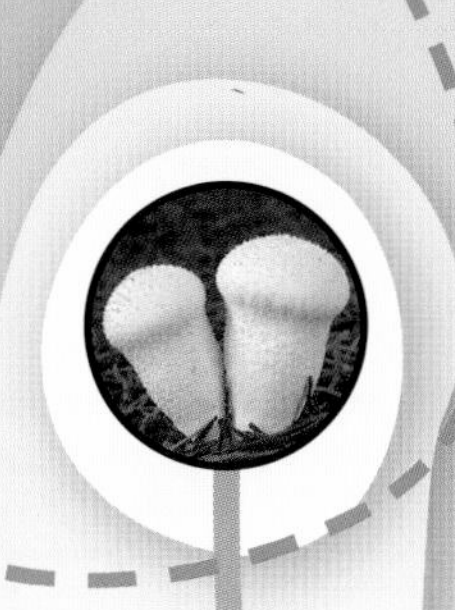

The upper levels of this system are called domains and kingdoms. There are three domains: *Archaea*, *Bacteria*, and *Eucarya*. The fungi belong to the *Eucarya*, which are characterized by having a nucleus in every cell.

Within the *Eucarya* the fungi are placed in their own kingdom, "*Fungi*," alongside the other kingdoms: *Chromista* (the brown algae and others), *Plantae* (plants), *Protozoa* (protozoans), and *Animalia* (animals). Of these, only the chromists and the plants can perform photosynthesis, whereas the fungi, animals, and most of the protozoa depend on digesting or decomposing organic matter created by the chromists and plants.

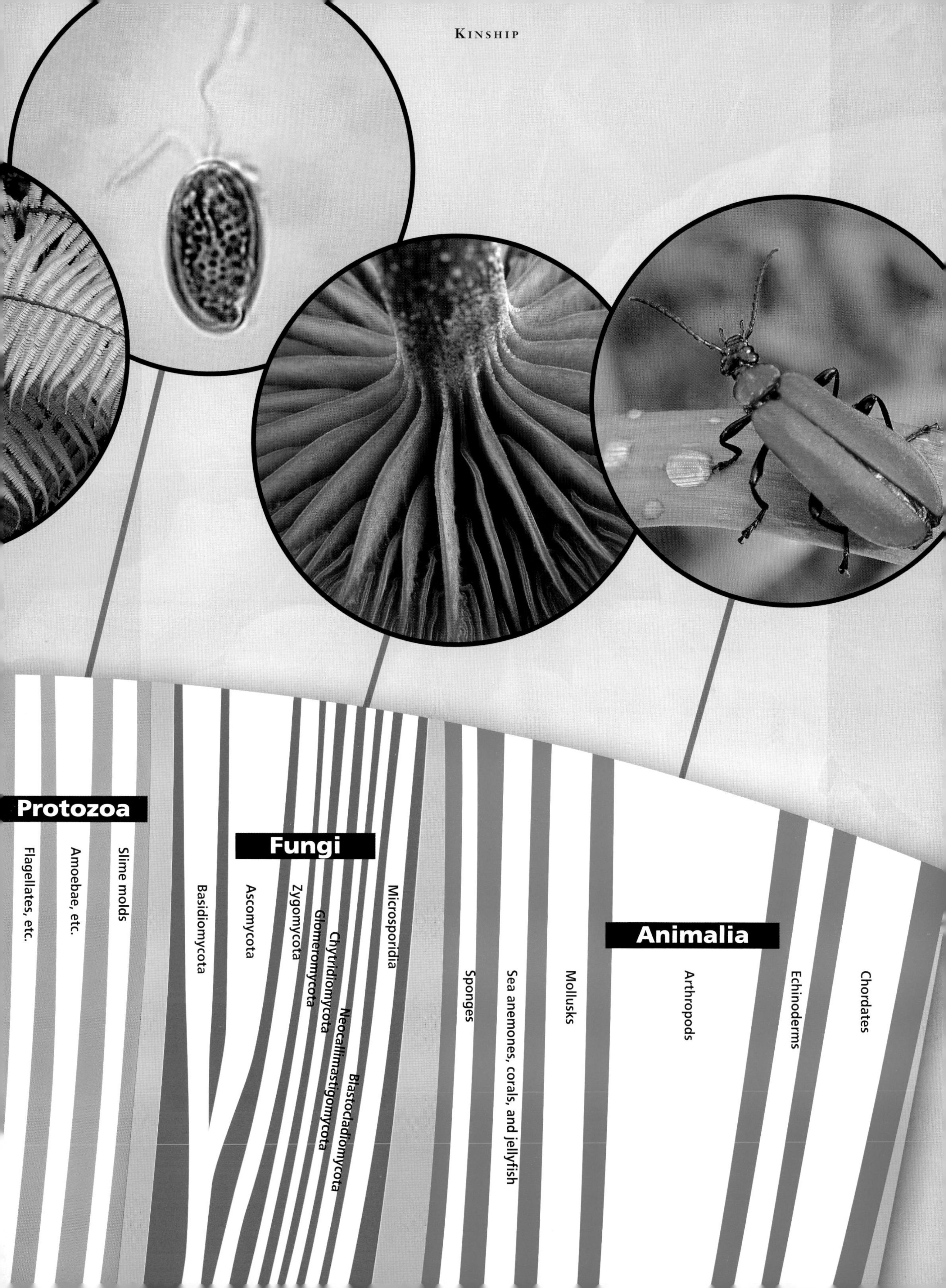
Protozoa
Flagellates, etc.
Amoebae, etc.
Slime molds
Fungi
Basidiomycota
Ascomycota
Zygomycota
Glomeromycota
Chytridiomycota
Neocallimastigomycota
Blastocladiomycota
Microsporidia
Animalia
Sponges
Sea anemones, corals, and jellyfish
Mollusks
Arthropods
Echinoderms
Chordates

The fungal kingdom

The fungal kingdom currently consists of eight major groups (phyla): *Microsporidia*, *Blastocladiomycota*, *Neocallimastigomycota*, *Chytridiomycota*, *Glomeromycota*, *Zygomycota*, *Ascomycota*, and *Basidiomycota* (the *Zygomycota* are a mishmash of groups that are not necessarily closely related). Only the *Ascomycota* and *Basidiomycota* (and the *Zygomycote* truffle genus *Endogone*) form fruiting bodies, and these are also by far the largest groups of fungi.

This schema represents the phyla and major classes of fungi. The approximate number of described species is given for each group.

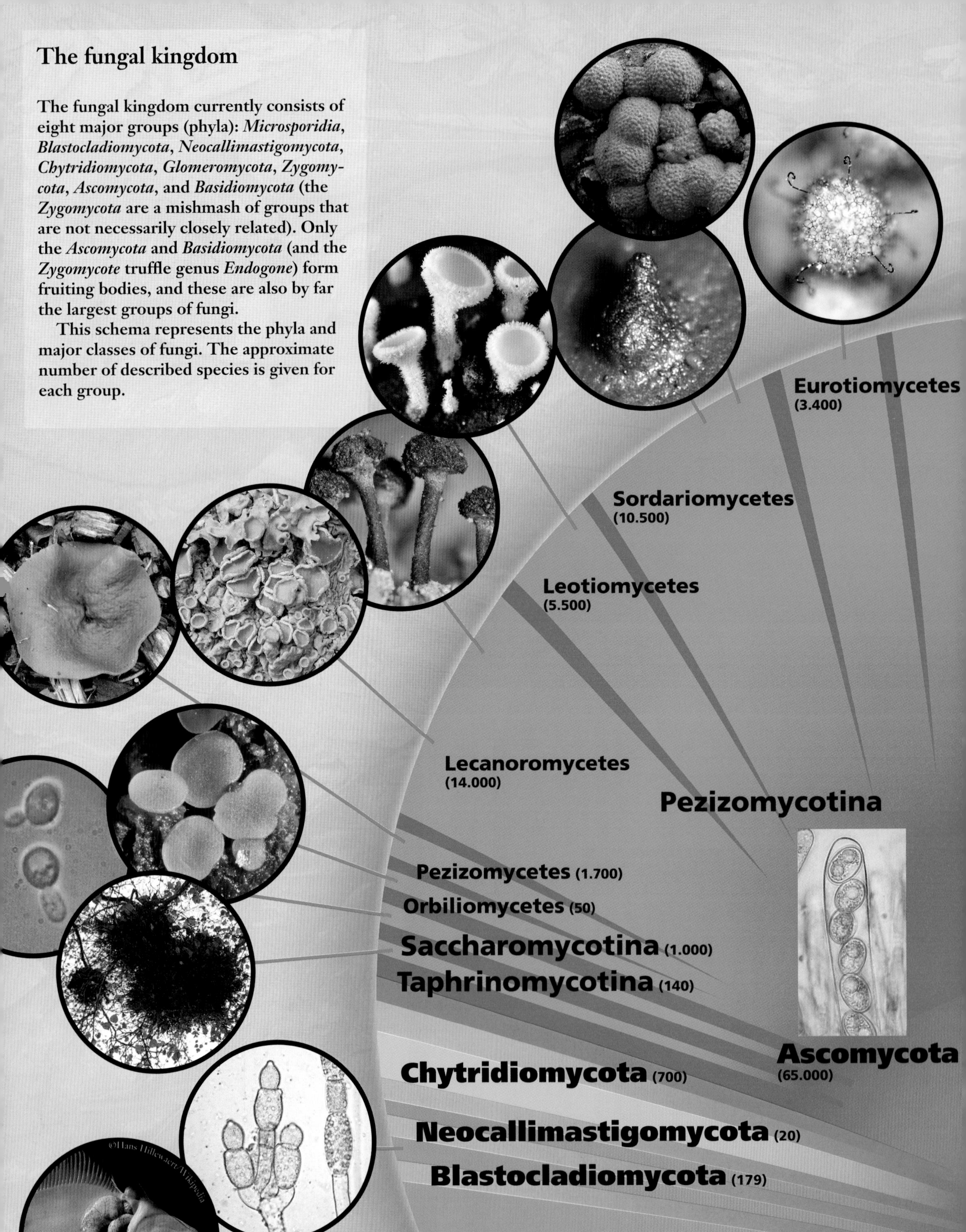

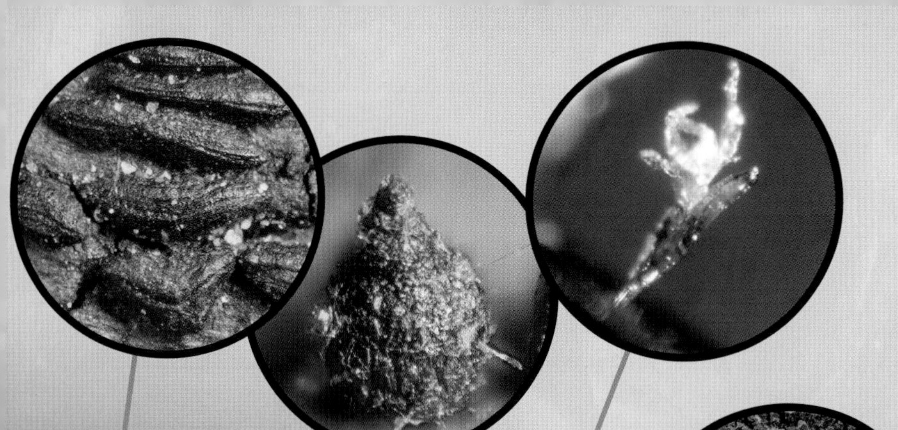

Dothideomycetes
(19.000)
Laboulbeniomycetes
(2.000)
Agaricomycetes
(21.000)
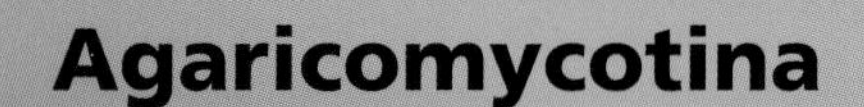
Agaricomycotina
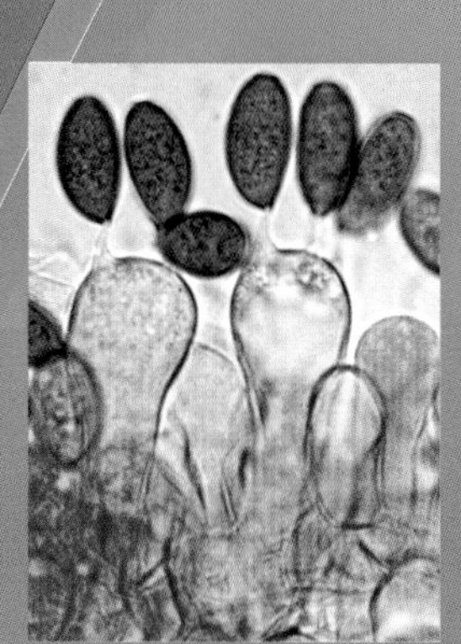
Dacrymycetes (100)
Tremellomycetes (400)
Ustilaginomycotina
(1.700)
Pucciniomycotina
(8.000)
Basidiomycota
(32.000)
"Zygomycota" (1.100)
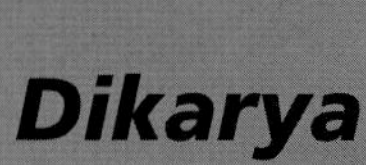
Dikarya

Glomeromycota (170)
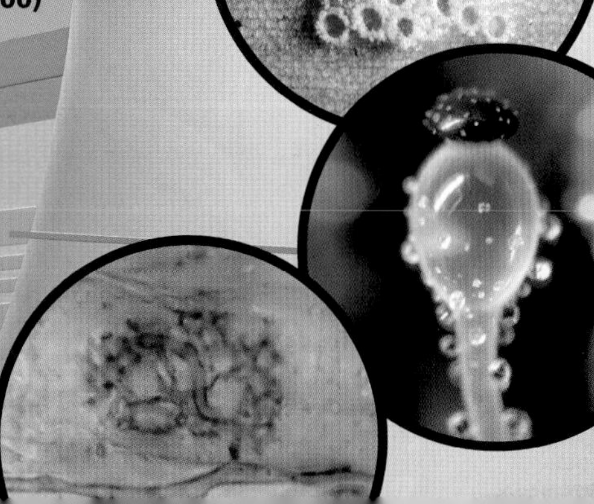

The perfect imperfects

Fungi have to disperse. When we find fungi in nature, we see mostly their fruiting bodies, which produce spores by sexual reproduction. The dispersal of sexually produced spores is one way to spread, but many fungi use a faster and more efficient way: they produce asexual spores. In some cases these spores are produced directly from the mycelium, but fungi usually form specialized hyphal structures to do the job. Although these are elegant and accomplish their purpose very efficiently, they were traditionally called "imperfect" (as opposed to the sexual state, called "perfect"). In more contemporary terms, we talk about the *asexual structures* of a fungus—or say that a fungus is anamorphic rather than teleomorphic.

Asexual states are probably best known for spoiling our food. When a piece of bread is left too long in the kitchen, asexual fungi are sure to arrive, forming white, blue-green, and black spots on the surface. If you blow on the bread, you will see clouds of spores dispersing into the air (but don't do this too often: the spores may be highly allergenic or even poisonous!)

Cakes left over from Christmas and strongly infected by asexual fungi.

Penicillium sp., Denmark (× 500)

At high magnification, the blue-green spots on the cakes show thousands of long chains of asexual spores (*conidia*) ready to be dispersed. The spores are produced from the top of erect hyphae called *conidiophores*.

Another very easy way to find asexual fungi is to look for stems and fruits of dead herbs. Here the surface is often partly covered by blackish or grayish "molds," mostly the asexual states of the Ascomycota.

Many molds (e.g., *Botrytis cinerea*) that grow naturally on herbs also cause serious trouble in agriculture. Anyone who grows strawberries knows how this fungus will destroy the fruits after a period of rain. Strangely, the rot can also be useful, since wine growers use grapes infected with *Botrytis cinerea* to produce a special sweet dessert wine. The fungus is then called *pourriture noble*—the noble rot.

Botrytis cinerea, Austria (× 15)

Botrytis cinerea, Denmark (× 5)

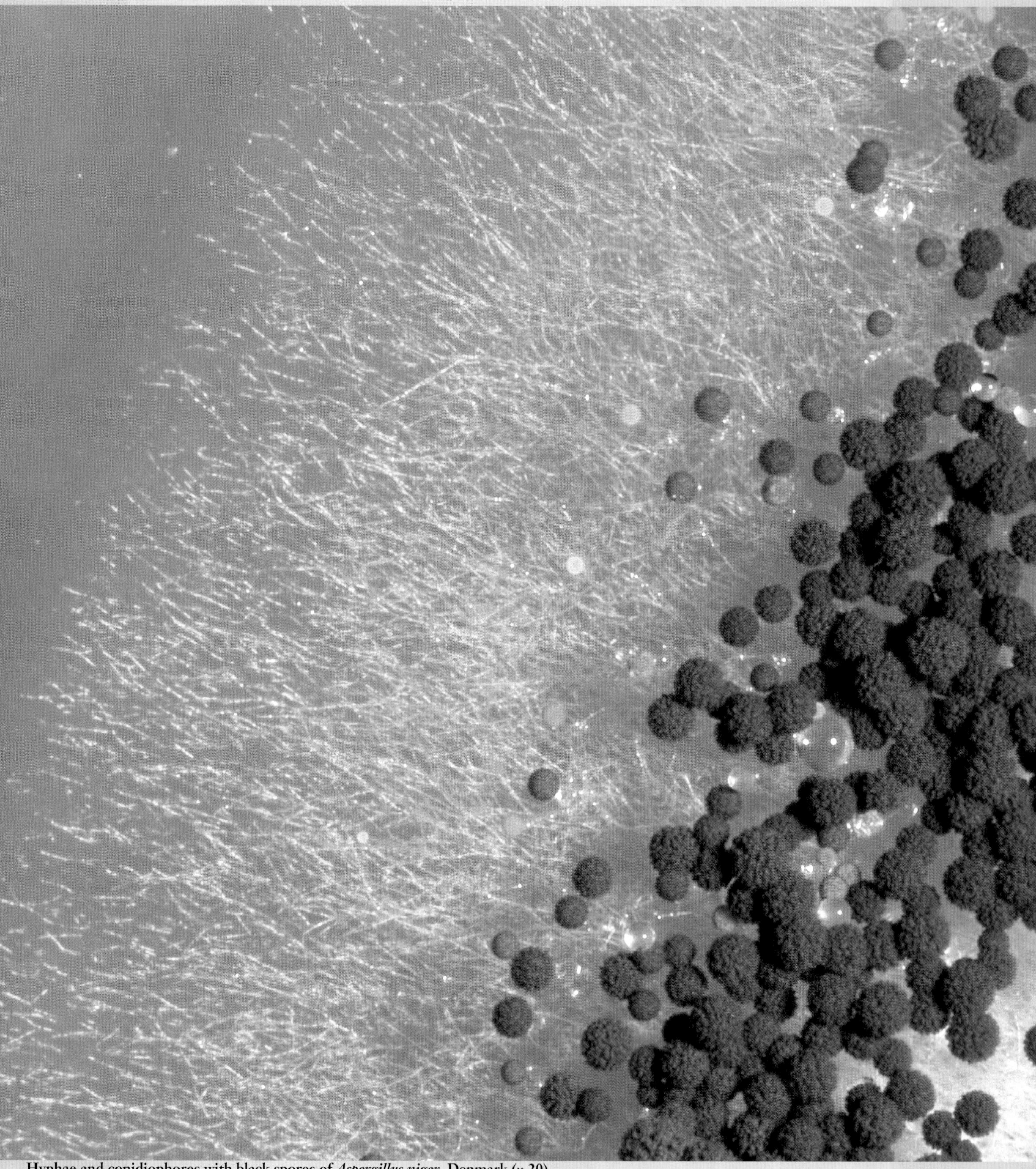

Hyphae and conidiophores with black spores of *Aspergillus niger*, Denmark (× 20)

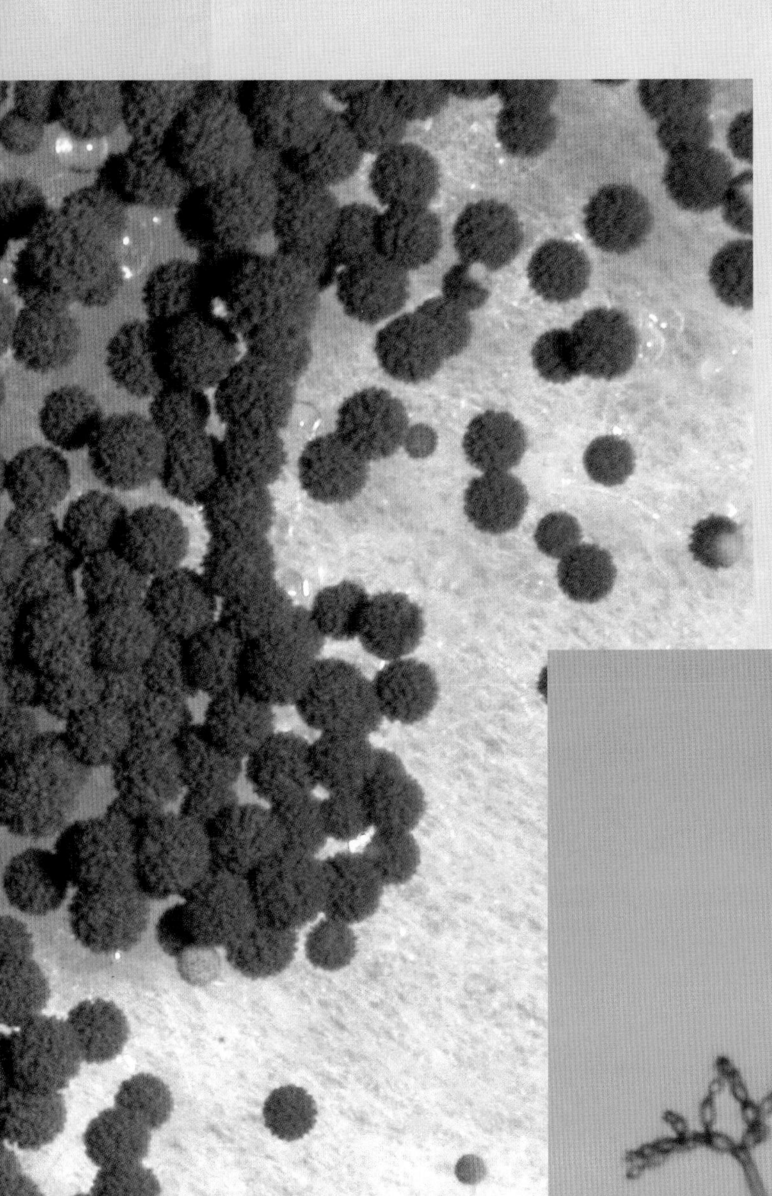

Asexual fungi form millions of spores. These are very small, often less than 1/100 of a millimeter, and are dispersed extremely easily by the wind. They are easily inhaled and some species are highly allergenic.

Species of *Alternaria*, *Aspergillus*, *Cladosporium*, *Mucor*, *Penicillium*, and *Rhizopus* are among the more common fungi in damp buildings and are thus some of the more potentially allergenic.

Cladosporium sp., Denmark (× 500)

All major groups of fungi can produce asexual states. Although the molds of the Ascomycota are the most common in buildings and on foods, the Zygomycota are also often present. These often form very characteristic structures that resemble small pins, with the spores contained in spherical heads (called *sporangia*).

Spore container of *Mucor* sp., Denmark (× 120)

Mucor sp. in the compost bin, Denmark

Rhizopus stolonifer, Denmark (× 100)

An especially elegant and efficient asexual fungus is *Rhizopus stolonifer*. This grows by offshoots, almost like strawberry plants. In this way the fungus can quickly move around, "shooting" out long hyphae that "root" in the substrate, forming a small bundle of hyphae with sporangia at the top.

Asexual fungi are major workhorses in the biotech industry. They can be grown in large tanks where they produce valuable compounds (called secondary metabolites). The products range from food additives (e.g., citric acid) to drugs and enzymes. Some of these compounds are produced naturally by the fungi, while others are genetically engineered to produce the desired product (see more on page 252).

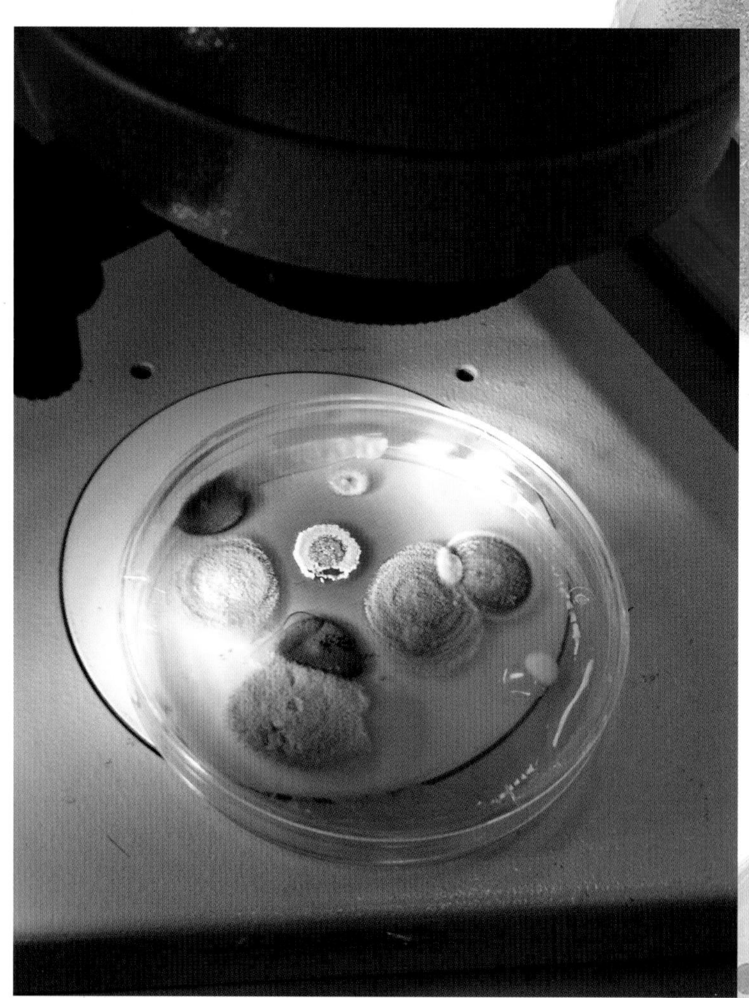

An agar plate under the dissecting microscope

Agar plates with asexual fungi from a biotech firm on display at a fungus exhibition

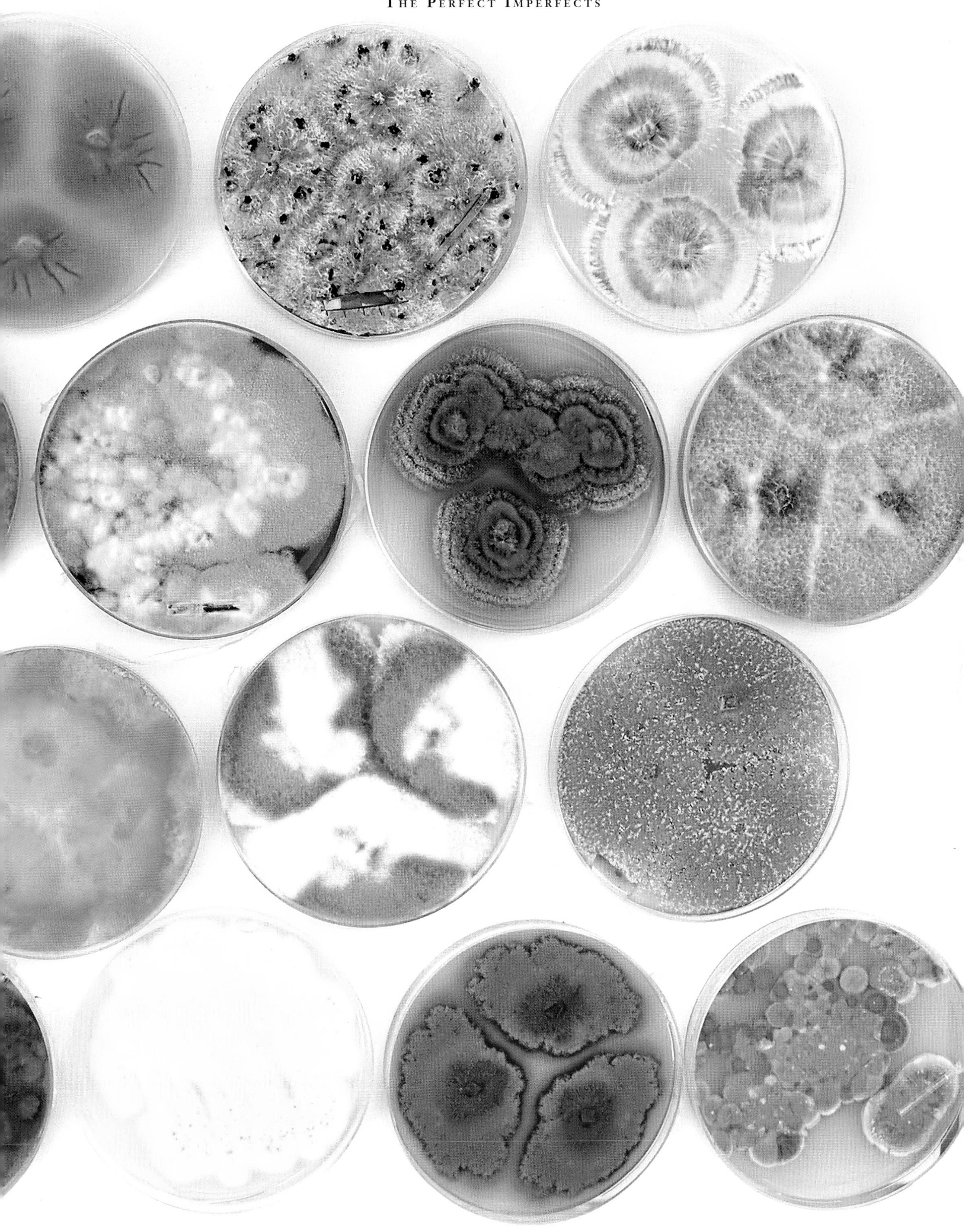

Fruiting bodies

To disperse themselves through sexual reproduction, fungi form *fruiting bodies* (also called carpophores, sporophores, ascocarps, or basidiocarps—or, more correctly, basidiomata and ascomata). Fruiting bodies are constructed of closely interwoven hyphae. Sometimes these hyphae resemble the hyphae of the mycelium, but they may also be specialized and found only in connection with fruiting bodies. For example, many fruiting bodies are partially built of strings of inflated cells (almost like pearl necklaces), and some perennial polypores are partly made of special thick-walled, skeletal hyphae.

Most fruiting bodies carry a palisade structure called a *hymenium*, where the reproductive cells are found and the sexual spores are formed. The hymenium on the discomycete below is on top of the fruiting body, whereas the hymenium on a typical basidiomycote is on a downward-pointing structure (for example, on the gills under the cap of an agaric).

A living fruiting body of *Ascobolus furfuraceus* with the dark hymenium at the top (× 20)

Section through *Ascobolus furfuraceus* with the hymenium at the top, Denmark (× 50)

Mycena pterigena, Denmark (× 250)

The *Mycena* above (one millimeter wide) has just emerged from its home in a dead fern stem. At this high magnification the construction of the cylindrical or sausage-shaped hyphae is easily seen.

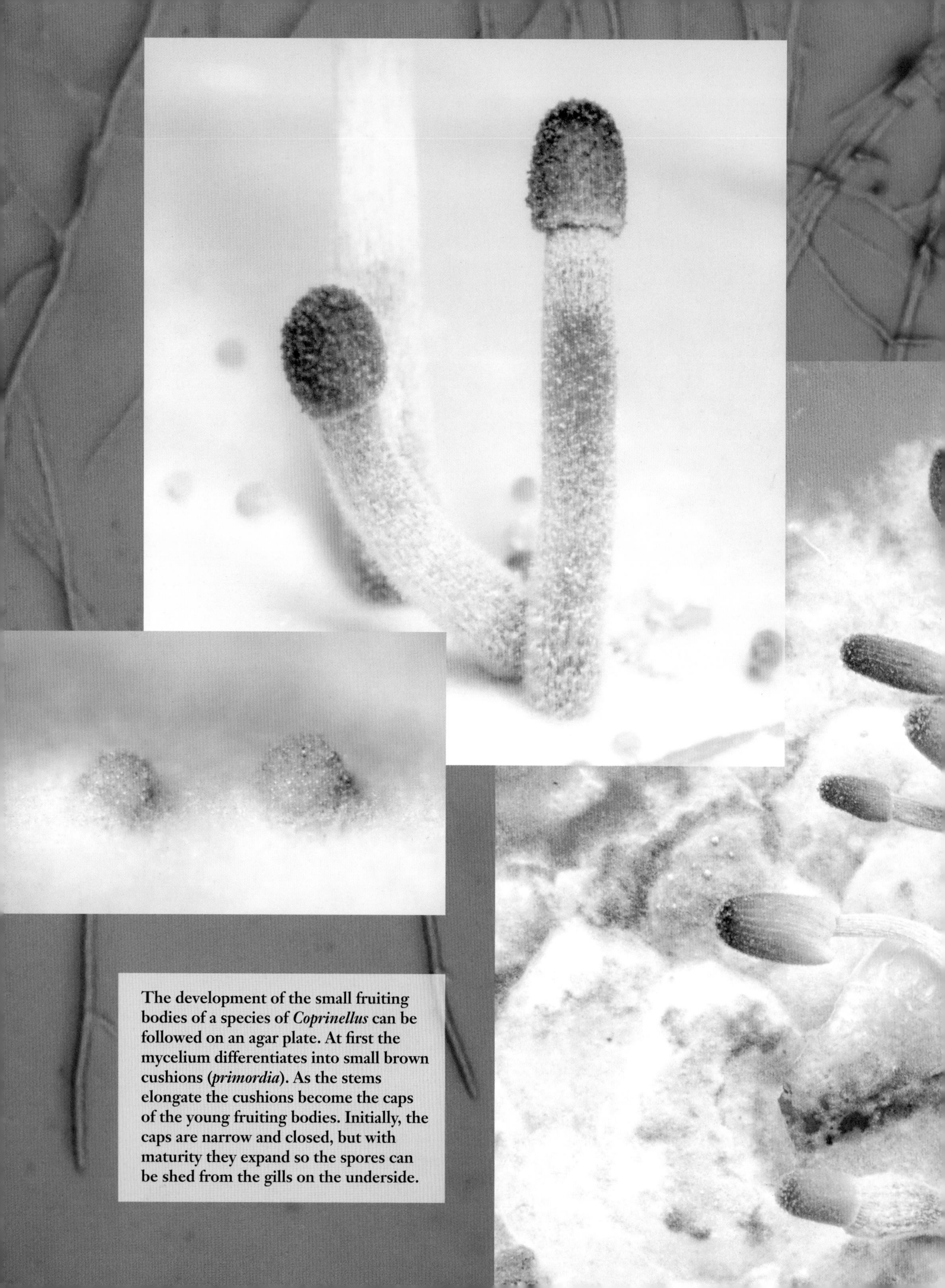

The development of the small fruiting bodies of a species of *Coprinellus* can be followed on an agar plate. At first the mycelium differentiates into small brown cushions (*primordia*). As the stems elongate the cushions become the caps of the young fruiting bodies. Initially, the caps are narrow and closed, but with maturity they expand so the spores can be shed from the gills on the underside.

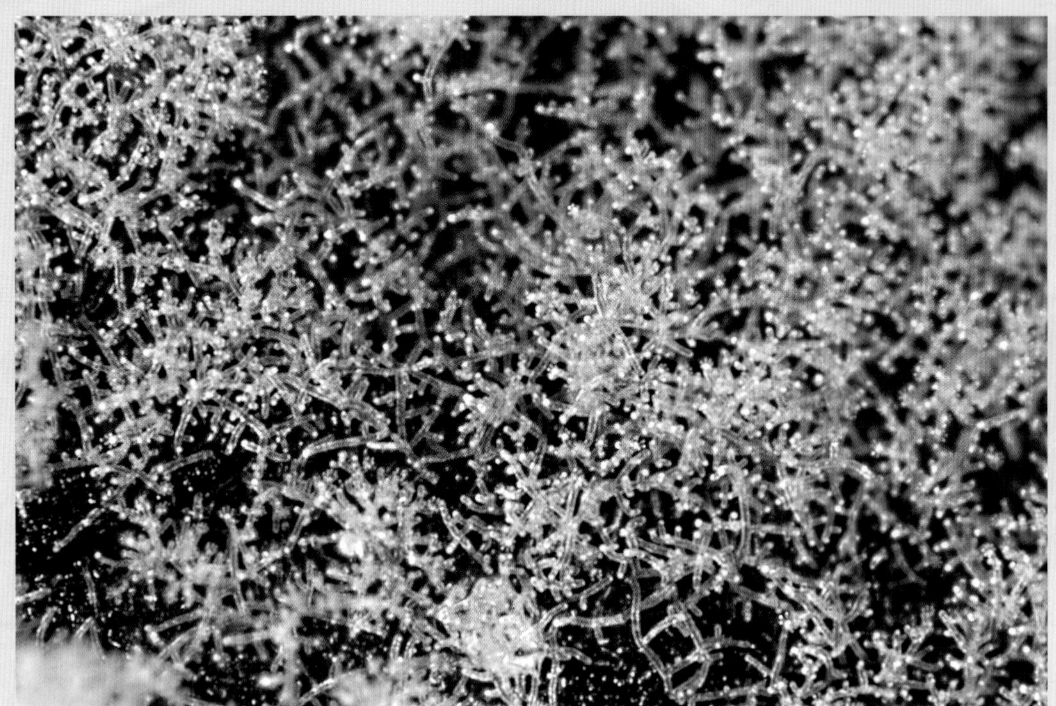

Botryobasidium pruinatum, Sweden (× 50)

Cheilymenia sp., Norway (× 20)

Cap of *Mycena capillaris*, Denmark (× 100)

The texture of fruiting bodies varies considerably, from hard as wood in the perennial polypores to loosely interwoven and almost moldlike in some flat (corticiaceous) fungi. The *Botryobasidium* at the top of the facing page is built of loose, branched hyphae with the reproductive basidia sitting freely at the hyphal ends.

The outer surface of fruiting bodies is often made of special cells like the pointed brown hairs (*setae*) on the outer side of a *Cheilymenia* (bottom left) or the clavate cap cuticle of a *Mycena* (above).

Parallel evolution—the disk/cup shape

When mycology emerged around 1800, fungi were divided according to the shape of their fruiting bodies. We now know that this does not reflect the evolutionary relationships of the species, because the same shapes have often evolved numerous times.

For example, the disk- to cup-shaped fruiting body is a universal shape that has been invented over and over again. It is found in at least seven different classes of fungi, in both the Ascomycota and the Basidiomycota. The group of fungi with disk-shaped fruiting bodies thus becomes a form group rather than a kinship group (see page 19).

Patellaria atrata, Denm… (× 20)

Mollisia ligni, Denmark (× 25)

Xanthoria parietina, Denmark (× 5)

Pseudographis pinicola, Sweden (× 15)

Peziza exogelatinosa, Denmark (× 1)

Orbilia coccinella, Denmark (× 15)

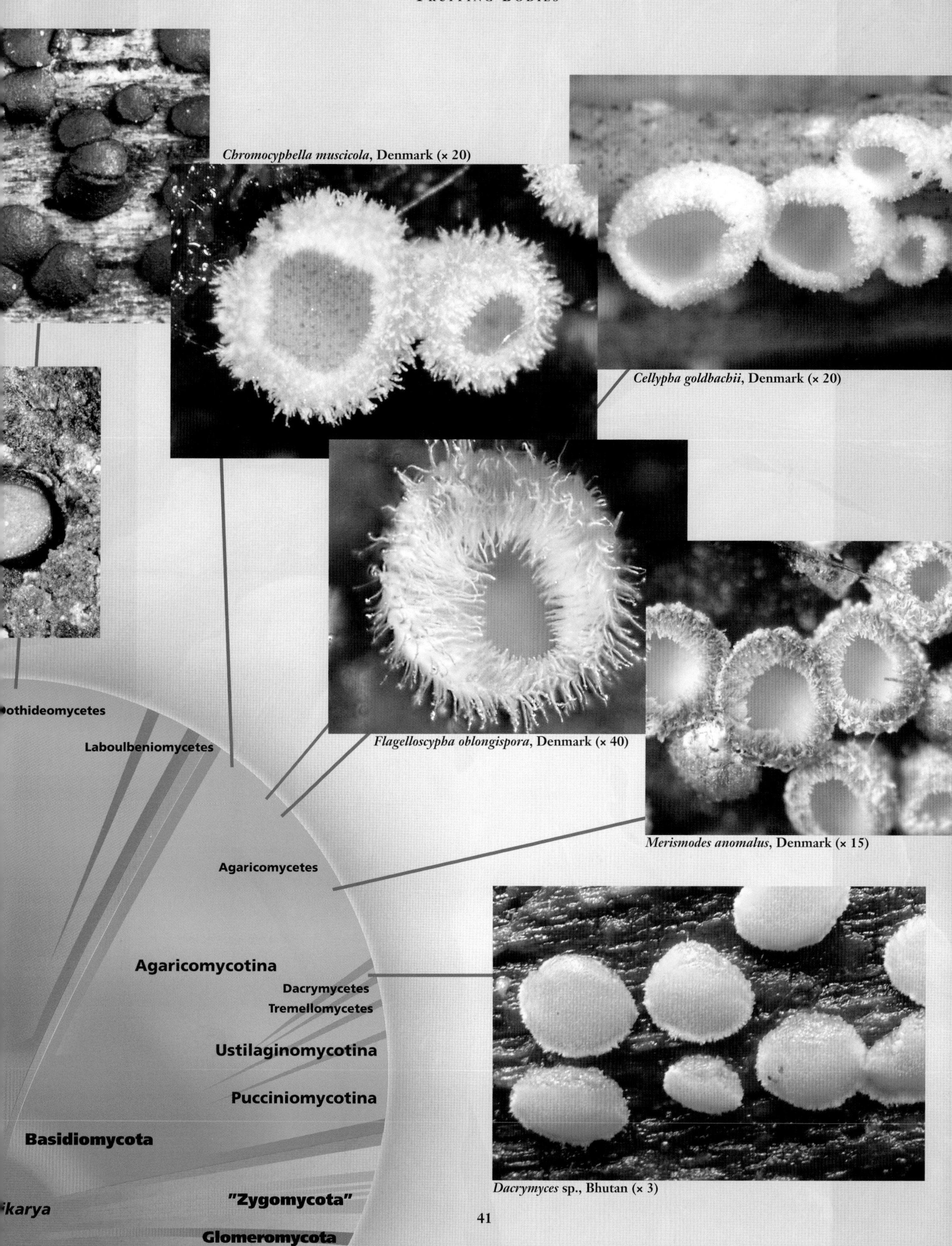

Chromocyphella muscicola, Denmark (× 20)

Cellypha goldbachii, Denmark (× 20)

Flagelloscypha oblongispora, Denmark (× 40)

Merismodes anomalus, Denmark (× 15)

Dacrymyces sp., Bhutan (× 3)

Parallel evolution—the club shape

As with the cup shape, the club-shaped fruiting body has also in the course of evolution been invented and reinvented in numerous groups of fungi. This should not be a surprise, as a club is one of the simplest shapes to produce.

Spathularia rufa, Norway (× 2)

Mitrula paludosa, Sweden (× 3)

Onygena equina, Northern Ireland (× 8)

Dibaeis sp., Bhutan (× 10)

Geoglossum cookeanum, Denmark (× 2)

Neolecta vitellina, Norway (TV, × 2)

Eurotiomycetes
Sordariomycetes
Leotiomycetes
Lecanoromycetes
Pezizomycotina
Pezizomycetes
Orbiliomycetes
Saccharomycotina
Taphrinomycotina
Chytridiomycota
Ascomycota
Neocallimastigomycota
Blastocladiomycota
Microsporidia

Clavaria argillacea, Denmark (× 2)

Clavicorona taxophila, Bhutan (× 5)

Clavariadelphus pistillaris, Denmark (× 0.5)

Calocera cornea, Denmark (× 5)

Eocronartium muscicola, Finland (× 5)

othideomycetes

Laboulbeniomycetes

Agaricomycetes

Agaricomycotina

Dacrymycetes

Tremellomycetes

Ustilaginomycotina

Pucciniomycotina

Basidiomycota

ikarya

"Zygomycota"

Glomeromycota

Fruiting body types

The following 144 pages illustrates the numerous types of fungal fruiting bodies. It is divided into the natural groups Ascomycota and Basidiomycota, whereas the subgroups (e.g., cup fungi, flask fungi, agarics, and polypores) are form groups and include species with a similar fruiting body type (see more on form groups on page 19).

Below is a graphic key to the major form groups treated in the book.

Key to the main form groups of fungi

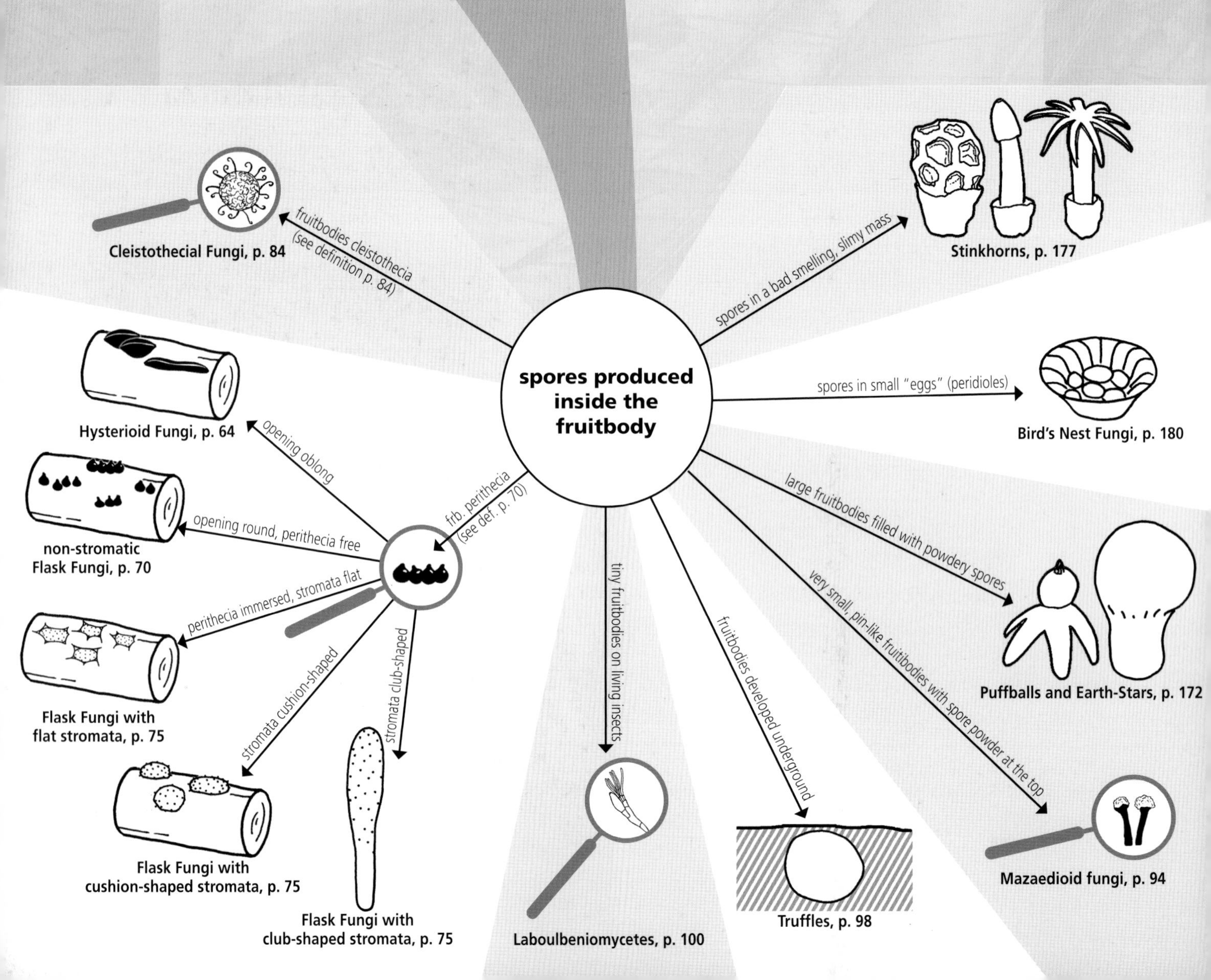

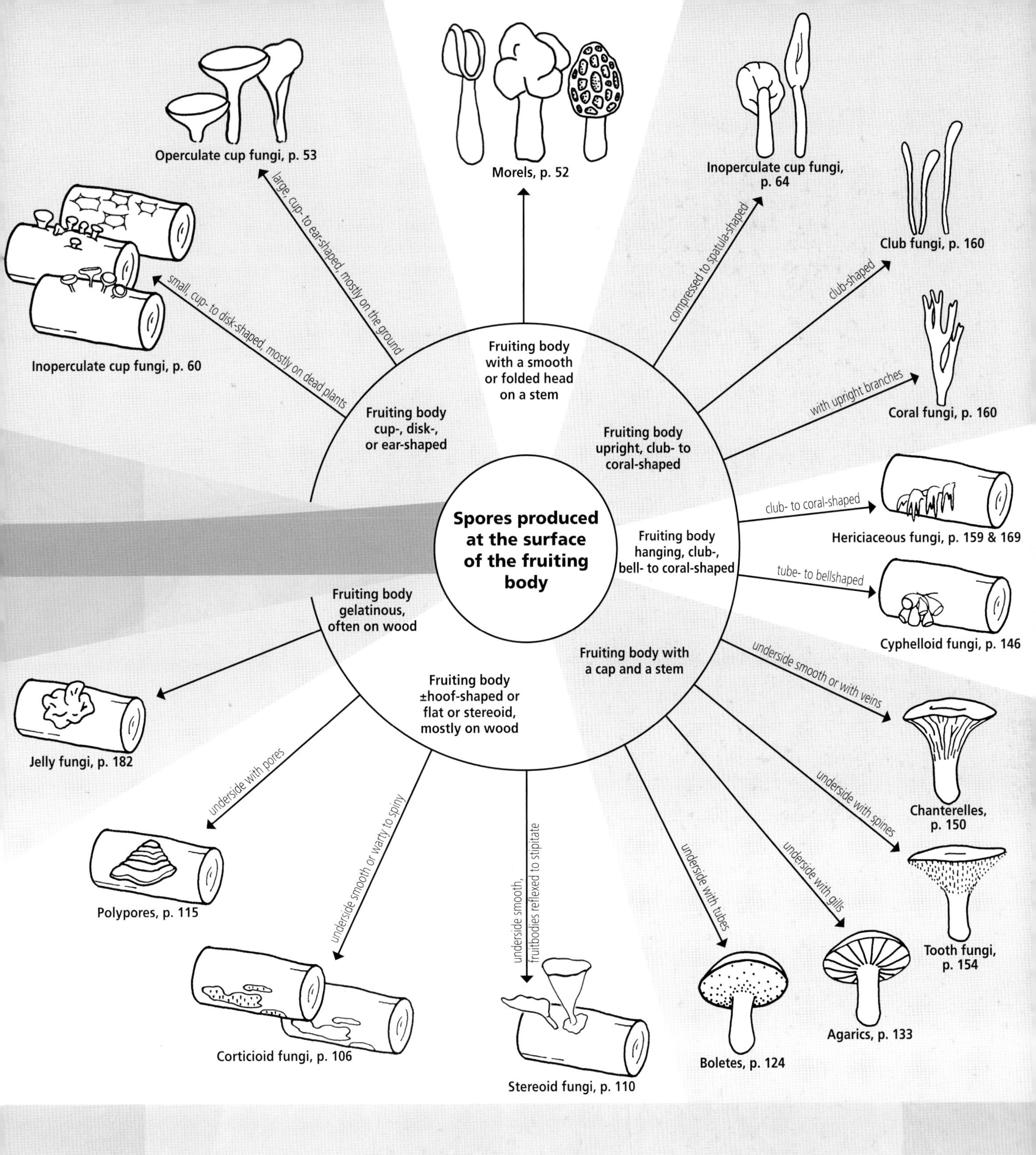

The above key for identifying the major groups of fungi is very simplistic. Some concepts, for example perithecia, are used in a much broader sense than normal. Some atypical members of form groups, for example stipitate polypores, will not key out, and some fungal shapes are not represented at all.

Unfortunately, precise identification of a fungus is very much a job for a specialist. For nonspecialists the first step may be to become familiar with the form groups in this key. This will be helpful when using the rest of the book or seeking further information.

The Ascomycota

The largest group of fungi is called the Ascomycota. The main, defining character for the Ascomycota is their production of sexual spores in cells called *asci*. The asci are normally found in a tissue called the *hymenium*. Here the asci sit parallel to each other, often intermixed with sterile hyphae called *paraphyses*. If the hymenium is colored, the pigment is often in the tips of the paraphyses. A single ascus usually contains eight spores, although the number may vary from one to several thousand.

Hymenium with asci and paraphyses of *Peziza varia*, Denmark (× 600)

The Ascomycota contain close to sixty-five thousand described species. The majority of these can form fruiting bodies. Some of these are minute, as small as one-tenth of a millimeter, whereas others, such as those of morels and truffles, are quite larger.

Historically, the Ascomycota were divided into formgroups based on the appearance of the fruiting bodies: species with open fruiting bodies were called discomycetes, and species with partially closed fruiting bodies pyrenomycetes. More popular names attached to these same form groups are cup fungi and flask fungi.

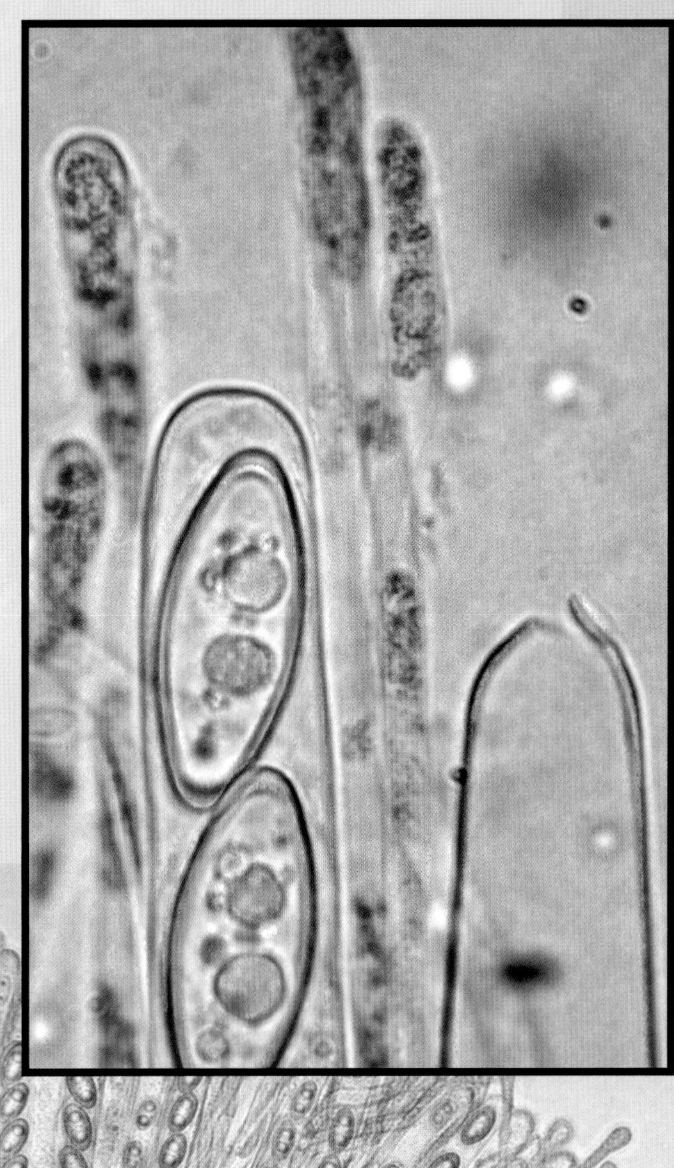

Asci and orange paraphyses of *Byssonectria terrestris*, Denmark (× 3500)

Aleuria aurantia, Denmark (× 450)

Gorgoniceps aridula with amyloid opening, Denmark (× 500)

Cucurbitaria obducens with thick-walled asci, Sweden (× 1000)

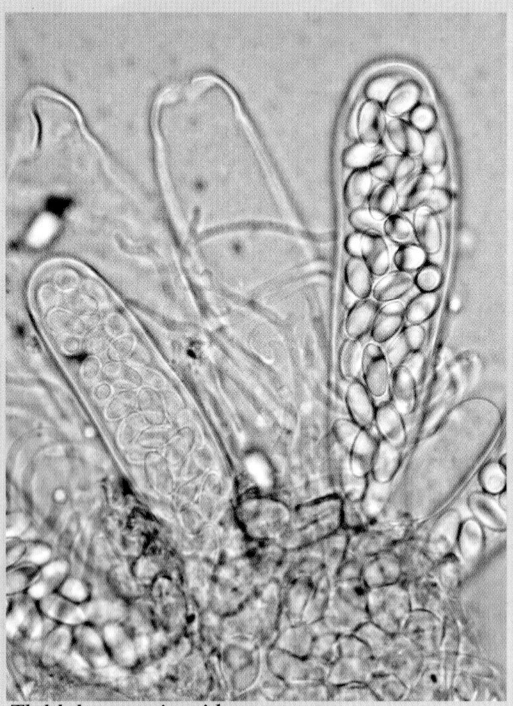

Thelebolus stercorius with many spores, Denmark (× 800)

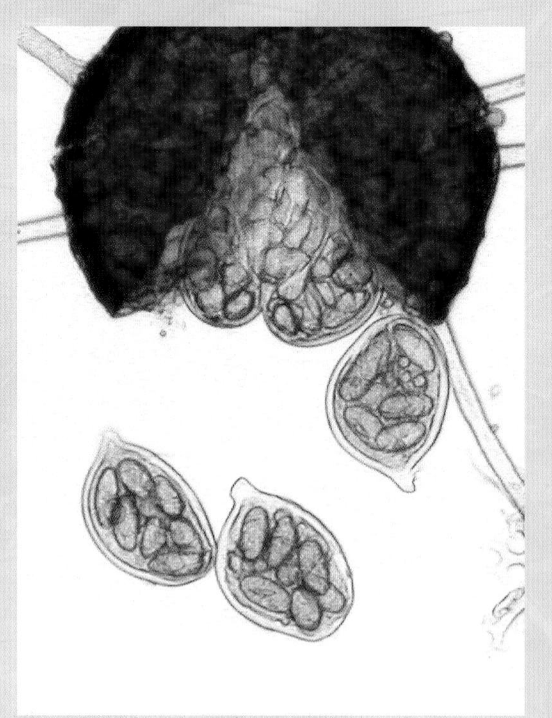

Erysiphe alphitoides without opening mechanism, Denmark (× 500)

The asci and spores of the Ascomycota vary greatly. The ascus opening in particular shows great variation, with thickened walls, lids, blue (*amyloid*) color reactions with iodine, or without any opening mechanism at all.

The Ascomycota may shoot the spores up into the air with great force. Often spores from many thousands of asci are released simultaneously, resulting in a cloud of spores arising from the hymenium accompanied by a hissing sound. This is called puffing and is generally initiated by a change in temperature above the fruiting body.

On the following 50 pages the fruiting body types of the major form groups of Ascomycota are illustrated.

Otidea umbrina (Denmark), before and during spore liberation (× 2)

Cup fungi

The cup fungi (or discomycetes) are characterized by open fruiting bodies, also called *apothecia*. The hymenium (the area containing the asci) is exposed and the asci can shoot their spores freely into the air. In the close-ups of fruiting bodies on this page you can see the asci protruding from the hymenium. In the lower two, you can even see the spores within the asci.

Cup fungi are not necessarily disk- or cup-shaped. In many cases they can be club-shaped, spatulate, or even have a folded hymenium sitting on top of a stem (facing page).

Lasiobolus papillatus with protruding asci, Denmark (× 50)

Two fruiting bodies of *Ascobolus albidus* with protruding asci and dark spores, Denmark (× 80)

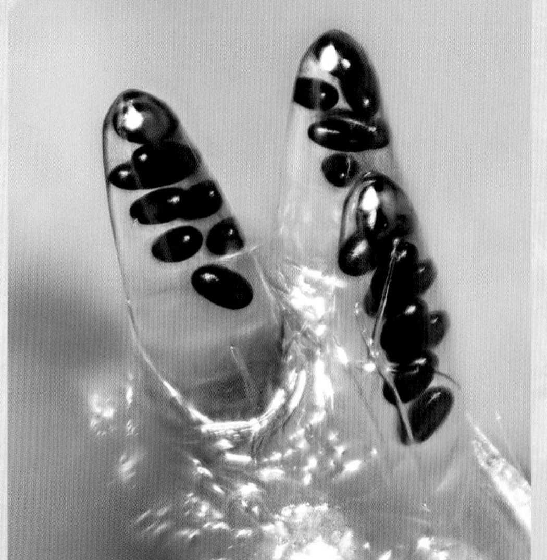

Giant asci of *Ascobolus immersus* with dark spores, Northern Ireland (× 180)

Helvella crispa, Denmark (× 3)

Operculate cup fungi

The operculate cup fungi are a subgroup of cup fungi characterized by asci that open by means of a small lid (an *operculum*, right). Most Ascomycota with large fruiting bodies belong here, but the group also contains species with extremely small fruiting bodies, especially among those growing on dung.

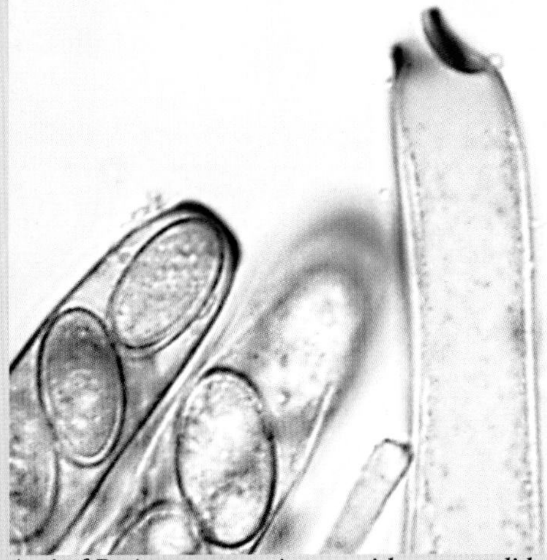

Asci of *Peziza arvernensis*, one with an open lid, Denmark (× 1200)

Cookeina tricholoma, Ecuador (× 3)

Morchella conica, Denmark (× 2)

The majority of the operculate cup fungi produce disk- or cup-shaped fruiting bodies. Here, the upper (inner) surface contains the hymenium, whereas the lower (outer) surface is sterile.

Pachyella babingtonii, Denmark (× 3)

Ramsbottomia asperior, Norway (× 8)

Aleuria aurantia, Denmark (× 2)

Peziza vesiculosa, Denmark (× 1)

Phillipsia domingensis, Ecuador (× 2)

Otidea concinna, Switzerland (× 1)

Peziza saniosa, Denmark (× 3)

Peziza gerardii, Denmark (× 4)

Scutellinia scutellata, Denmark (× 15)

Leucoscypha leucotricha, Denmark (× 20)

Parascutellinia carneosanguinea, Norway (× 8)

Trichophaea hybrida, Denmark (× 8)

Dezmazierella acicola, Denmark (× 25)

Leucoscypha leucotricha, Denmark (× 20)

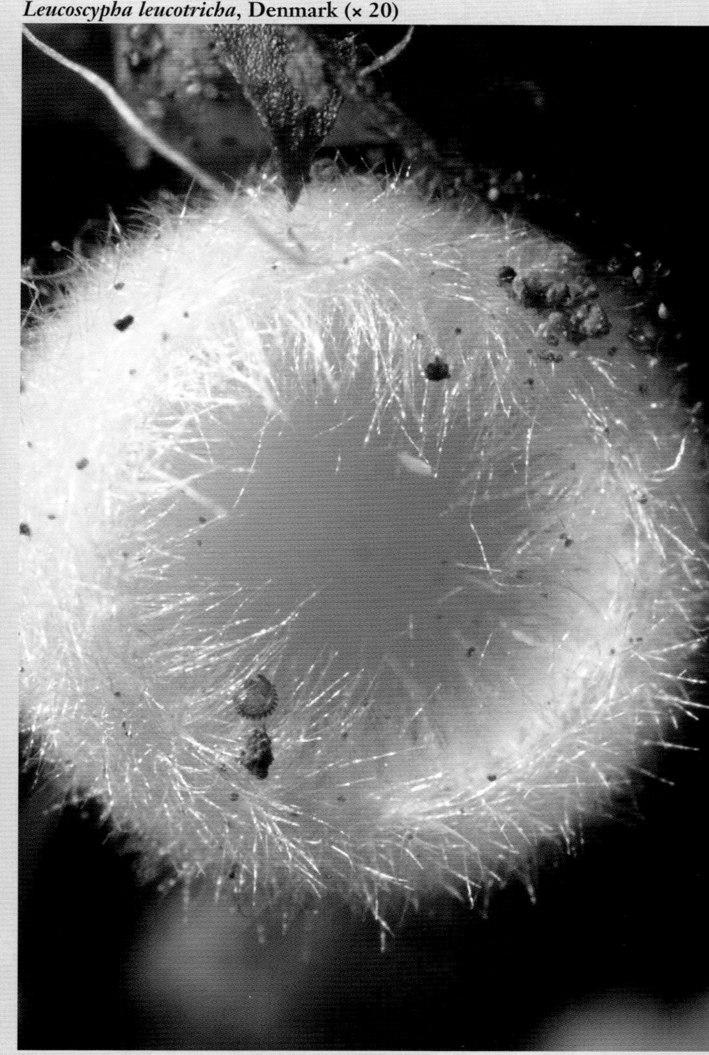

Plectania melastoma, Denmark (× 12)

In many operculate cup fungi, the sterile outer surface of the fruiting bodies is covered by hairs of various lengths. In some instances even the hymenium has hairs protruding from the surface (upper left). The hairs may provide protection against small gnawing animals.

Melastiza cornubiensis, Denmark (× 25)

Helvella bulbosa, Denmark (JV, × 2)

Helvella lacunosa, Norway (× 2)

Spores are powerfully liberated from cup fungi, and this usually succeeds in getting the spores airborne. But for some species this is not enough: they have developed tall or stalked fruiting bodies to thrust the spores even farther into the air.

Otidea onotica
Denmark (× 4)

Inoperculate cup fungi

The inoperculate cup fungi are characterized by open fruiting bodies (*apothecia*) with asci that open without a lid (right). They mostly form small fruiting bodies and often decompose woody plants or herbs.

Because many species are bound to specific host plants, the number of species worldwide is expected to be very high.

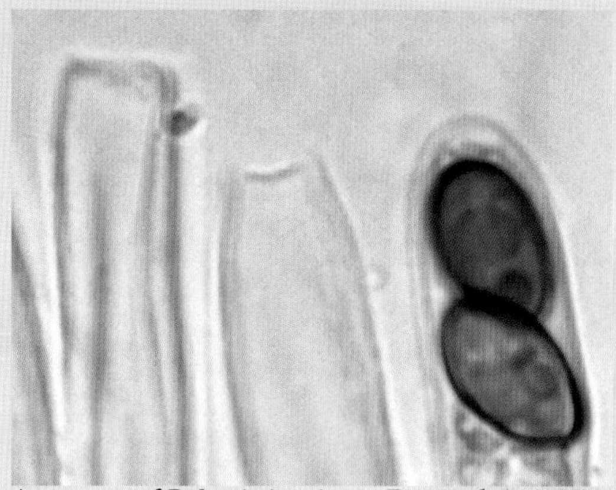

Ascus tops of *Bulgaria inquinans*, Denmark (× 1400)

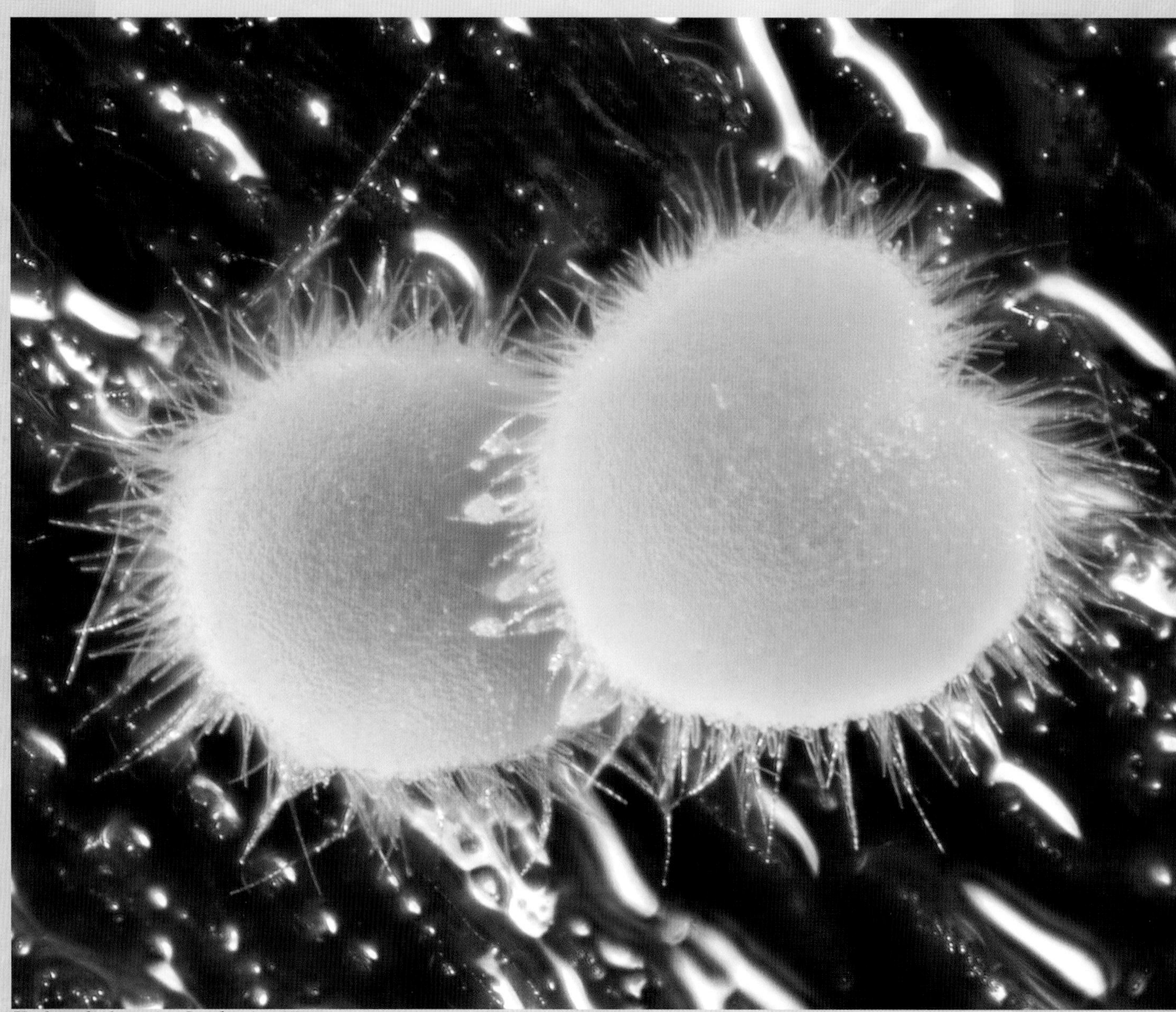

Hyaloscypha leuconica, Sweden (× 150)

Roseodiscus rhodoleucus, Denmark (× 7)

Lachnum imbecile colored by orange paraphyses, Sweden (× 120)

Most inoperculate cup fungi form disk- or cup-shaped fruiting bodies similar to those of the operculate fungi: the inner surface contains the hymenium and the outer surface is sterile. The sterile side may be smooth or covered with hairs.

Orbilia delicatula, Denmark (× 8)

Bisporella citrina, Denmark (× 6)

Mollisia cinerea, Denmark (× 8)

Ascocoryne sarcoides, Denmark (× 3)

Chlorociboria aeruginascens, Denmark (× 3)

Crocicreas coronatum, Denmark (× 6)

Lachnum pudibundum, Denmark (× 25)

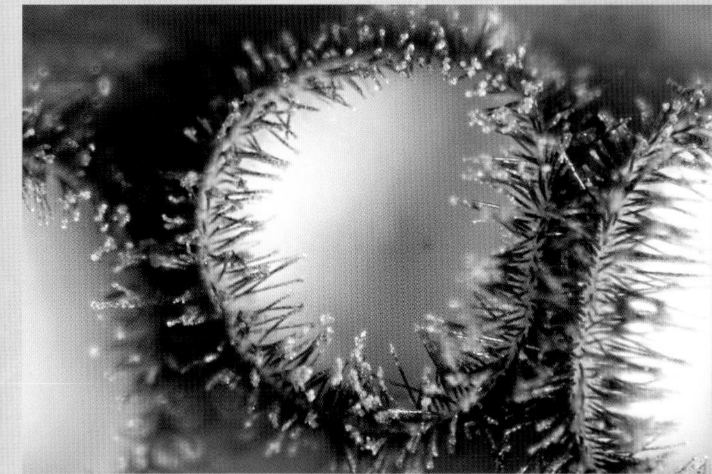
Trichopezizella relicina, Austria (× 50)

Trichopeziza leucophaea, Denmark (× 25)

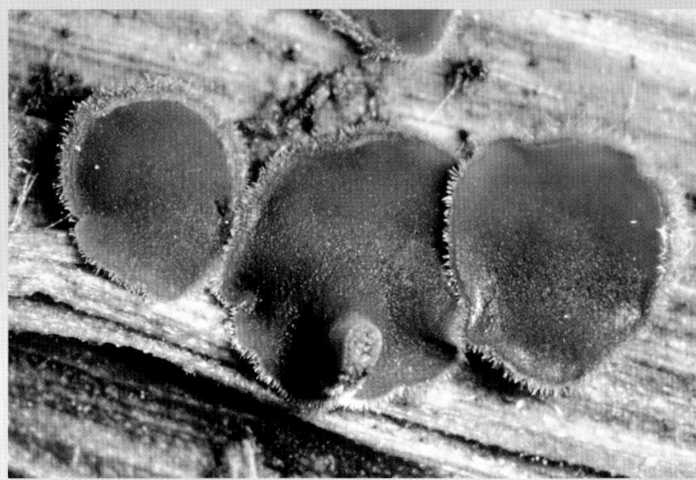
Dennisiodiscus prasinus, Denmark (× 20)

Perrotia flammea, Sweden (× 50)

Lachnellula suecica, Denmark (× 15)

Mitrula paludosa, Norway (× 1.5)

Microglossum viride, Denmark (TF, × 2)

Geoglossum starbaeckii, Denmark (× 3)

Geoglossum glutinosum, Denmark (JV, × 2)

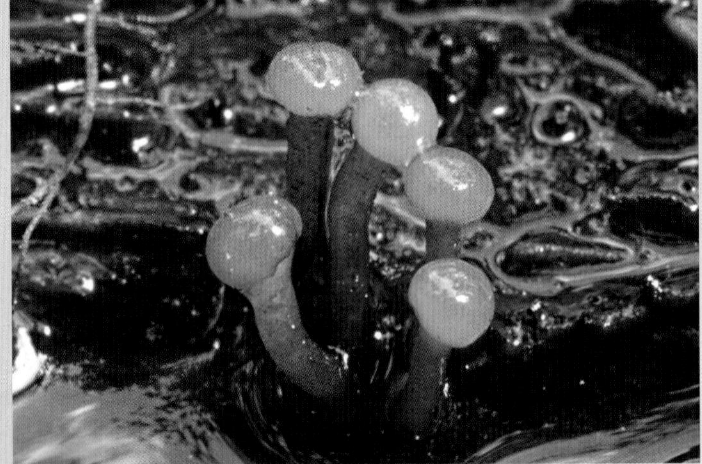
Vibrissea truncorum, Norway (× 4)

Spathularia flavida, Denmark (× 2)

A significent number of inoperculate cup fungi have evolved long stems that lift the hymenium into the air, in some instances transforming the whole fruiting body into a small club.

Leotia lubrica, Denmark (× 5)

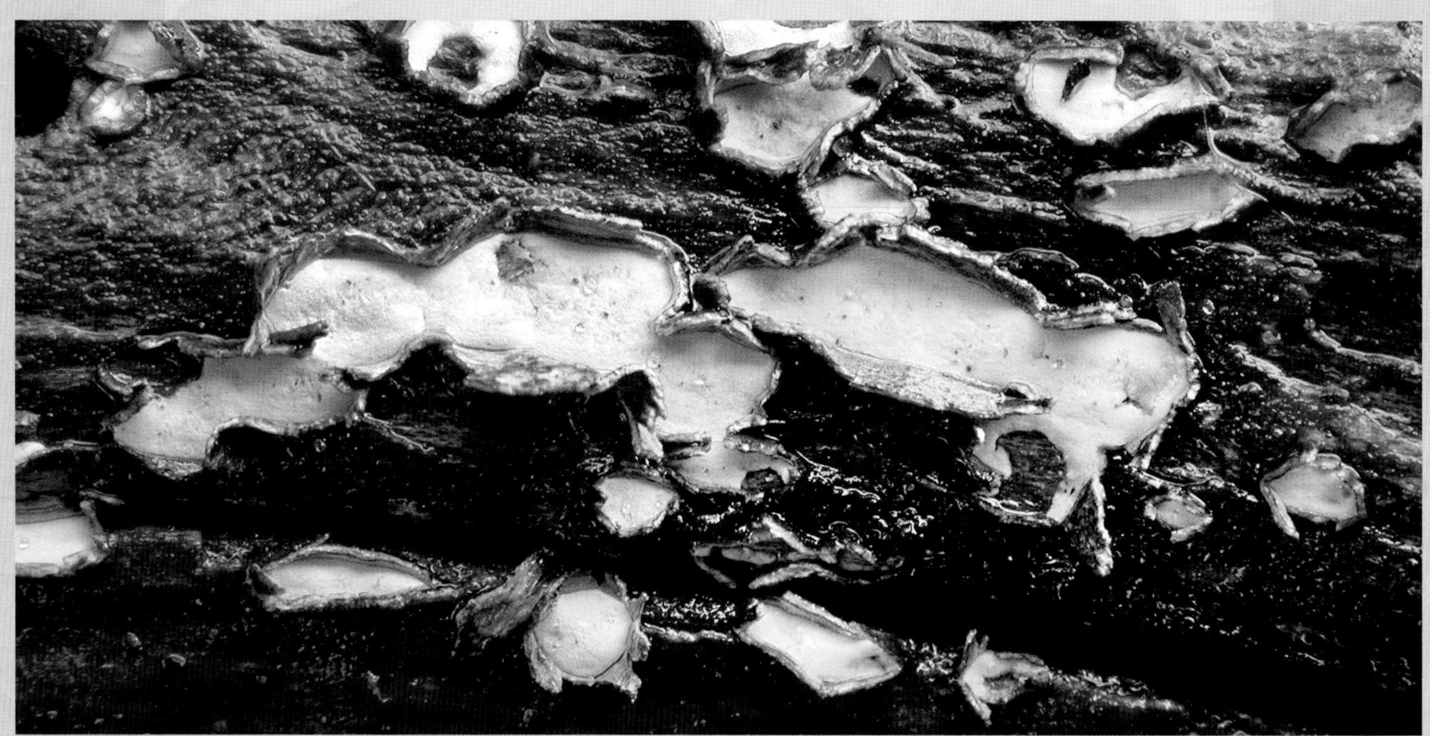

Propolis farinosa, Denmark (× 8)

Colpoma crispum, Sweden (× 80)

One of the constant challenges facing small fungi is that of surviving drought long enough to spread their spores. Some cup fungi have solved this by partly immersing their fruiting bodies in the substrate. They may then swell in moist weather, forcing the fruiting body open and exposing the hymenium. The most elegant have even developed a small lid to open and close in response to moisture.

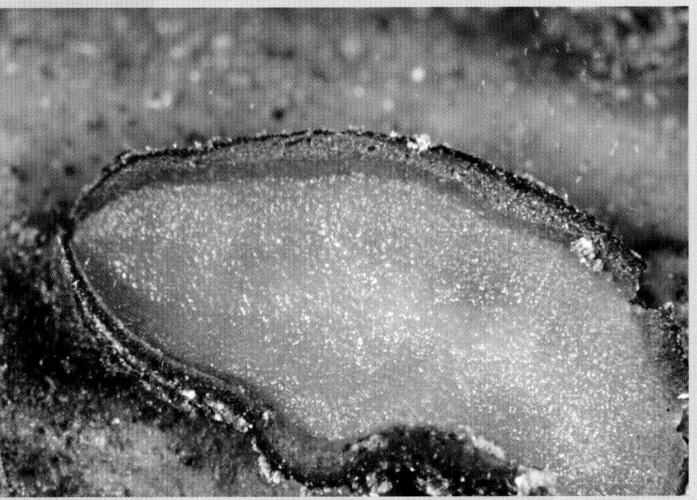
Colpoma quercinum under wet conditions, Denmark (× 20)

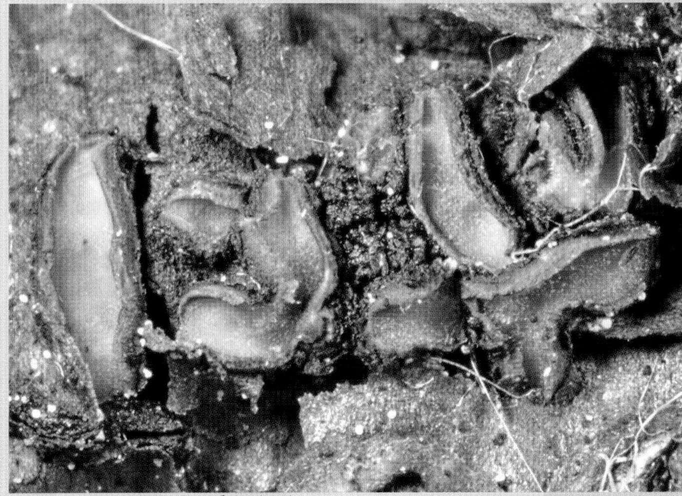
Colpoma juniperi, Sweden (× 8)

Lophodermium conigerum, Denmark (× 30)

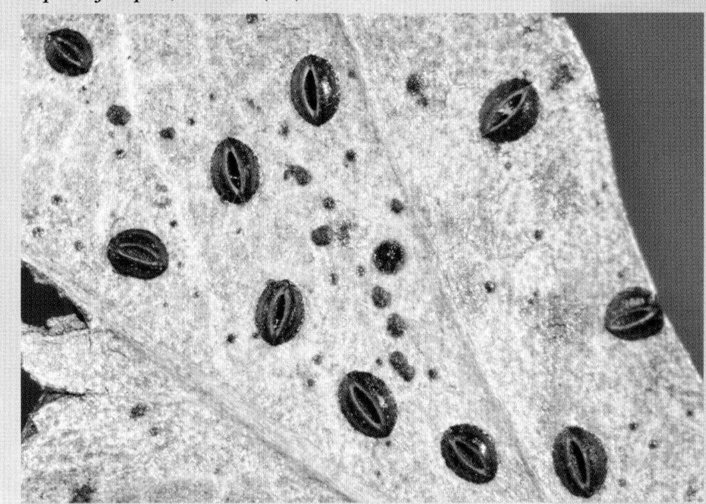
Lophodermium foliicola, Denmark (× 10)

Rhytisma acerina with exposed hymenia, Denmark (× 4)

Trochila ilicina with lid open, England (× 80)

Hysterioid fungi

In their attempt to preserve moisture and tolerate drought, some fungi have evolved even further than the immersed cup fungi on the previous pages. The hysterioid fungi and the flask fungi have both evolved a hard outer crust and a partially closed fruiting body. Whereas the flask fungi open through a small pore (the *ostiole*), the hysterioid fungi open through a crack or slit.

The hymenium is never exposed, so the asci have to stretch to reach the opening crack before releasing the spores.

Hysterobrevium mori, Denmark (× 15)

Lophium mytilinum, Sweden (× 70)

Glyphium elatum, Denmark (× 60)

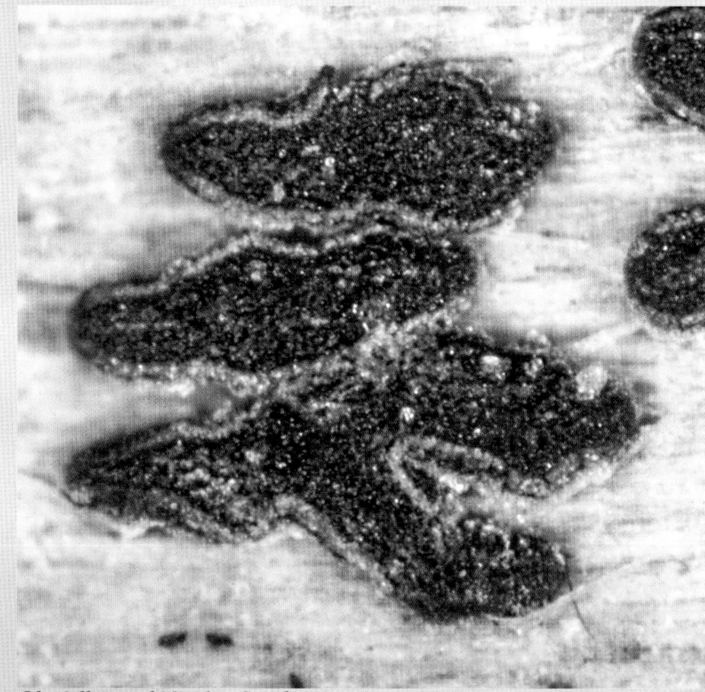

Gloniella graphidioidea, Sweden (× 40)

Hysterium angustatum, Denmark (× 40)

Flask fungi

The masters of drought resistance are the flask fungi (or pyrenomycetes). With their almost-closed fruiting bodies (called *perithecia*) and often-hard exterior, they can tolerate drought and direct sun exposure for days.

During spore release, one ascus stretches to the top opening (the *ostiole*) and shoots its spores. Because of this, the fruiting bodies are normally no longer than an ascus can stretch: less than two millimeters.

In some cases many fruiting bodies are placed inside a common tissue called a *stroma*, thus forming large structures that are more easily seen (facing page, bottom).

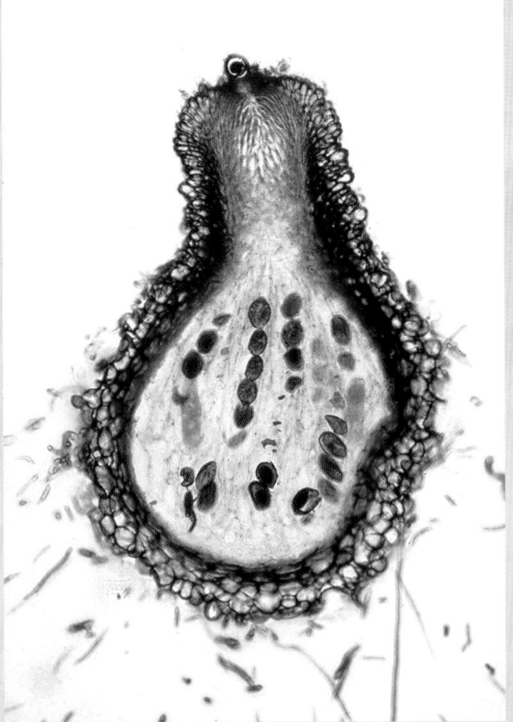

Section through *Sordaria fimicola* with asci and dark spores inside, Denmark (× 230)

Fruiting bodies of *Neonectria punicea* with masses of white spores caught in the opening, Denmark (× 80)

Acrospermum compressum shooting long, filiform spores, Denmark (× 50)

Camarops polysperma shooting small spores from numerous fruiting bodies (the small protruding warts) immersed in a stroma, Denmark (× 40)

Rosellinia aquila with spores on top, Denmark (× 40)

Barbatosphaeria barbarostris, Denmark (× 30)

Echinosphaeria canescens , Denmark (× 20)

Cercophora sp., Denmark (× 40)

Leptosphaeria doliolum, Denmark (× 40)

Acanthophiobolus helicosporus, Denmark (× 30)

Although flask fungi may seem simple in construction, there is room for endless variation. The fruiting bodies can have quite diverse shapes, from spherical or pear-shaped to oblong, and they can sometimes have a long beak or neck. The outer surface can be smooth or covered with hairs.

Annulohypoxylon cohaerens, Denmark (× 2)

Annulohypoxylon multiforme, Denmark (× 4)

Hypoxylon fragiforme, Denmark (× 1)

Diatrype disciformis, Denmark (× 2)

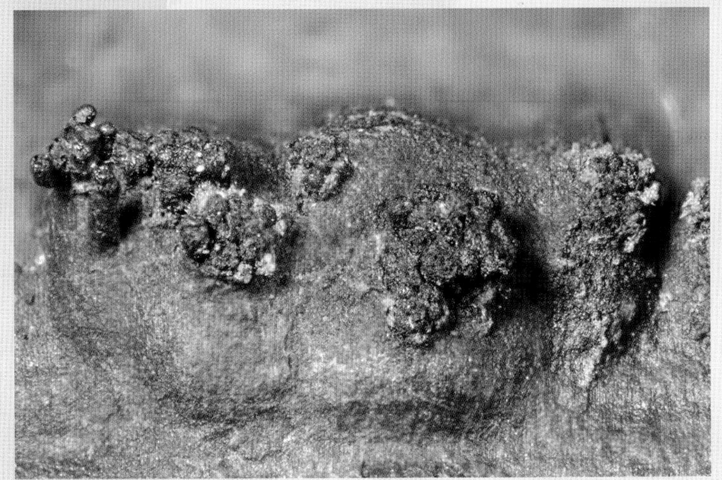
Valsa sp., Denmark (× 20)

Diaporthe aristata, Sweden (× 15)

When the fruiting bodies are collected inside a common tissue (or *stroma*), the structure becomes larger and easily visible. Here as well, great variation is possible.

The form group is collectively known as the stromatic flask fungi.

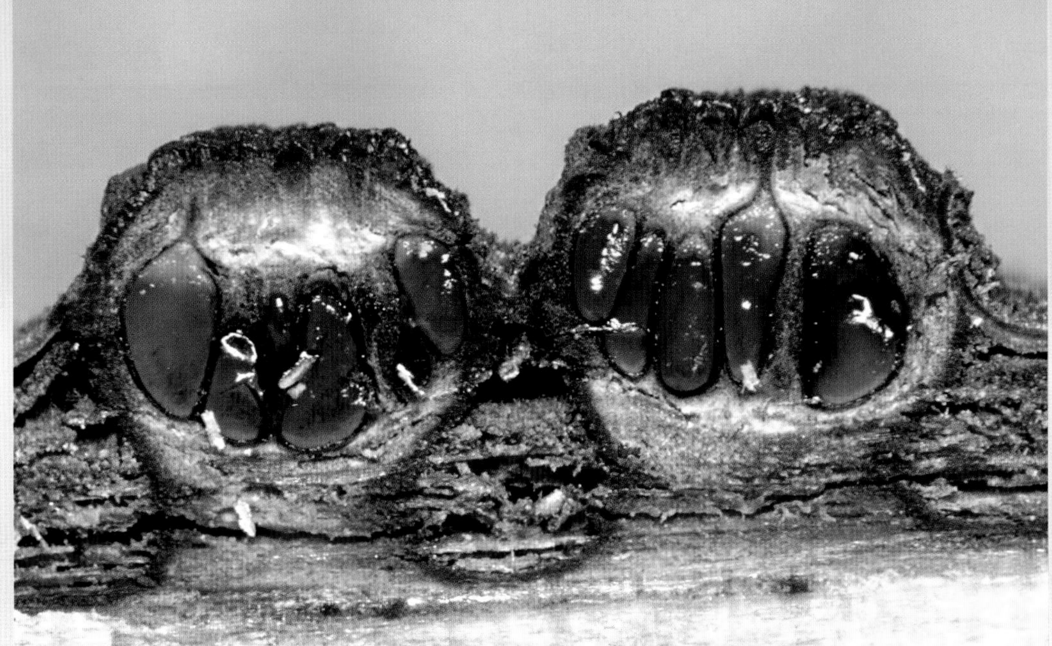

Entire and sectioned stromata of *Diatrypella quercina*, Denmark (× 4 and 12)

The stromatic flask fungi may have numerous fruiting bodies gathered inside the stroma. The openings of the fruiting bodies reach the surface and give the stroma a spotted appearance. The single fruiting bodies (*perithecia*) are easily seen if you make a section through the structure.

Xylaria telfairii, Ecuador (× 3)

Camillea patouillardii, Ecuador (× 4)

Camillea sulcata, Ecuador (× 4)

Camillea cyclops, Ecuador (× 6)

Camillea stellata, Ecuador (× 4)

Camillea mucronata, Ecuador (× 6)

Camillea leprieurii, Ecuador (× 1)

The family Xylariaceae contains mostly stromatic flask fungi. The species exhibit a great diversity, especially in the tropics, where they produce some of the most exotic shapes of any fungi.

Hypoxylon nicaraguense and *Rosellinia* sp., Burkina Faso (× 1)

Thamnomyces dendroidea, Ecuador (TL, × 1.5)

The fruiting bodies of the flask fungi are not always hard little brown or black blobs. In the subgroup of hypocrealean fungi, many species are rather soft and brightly colored. Some of these species are parasitic on plants, insects, or other fungi.

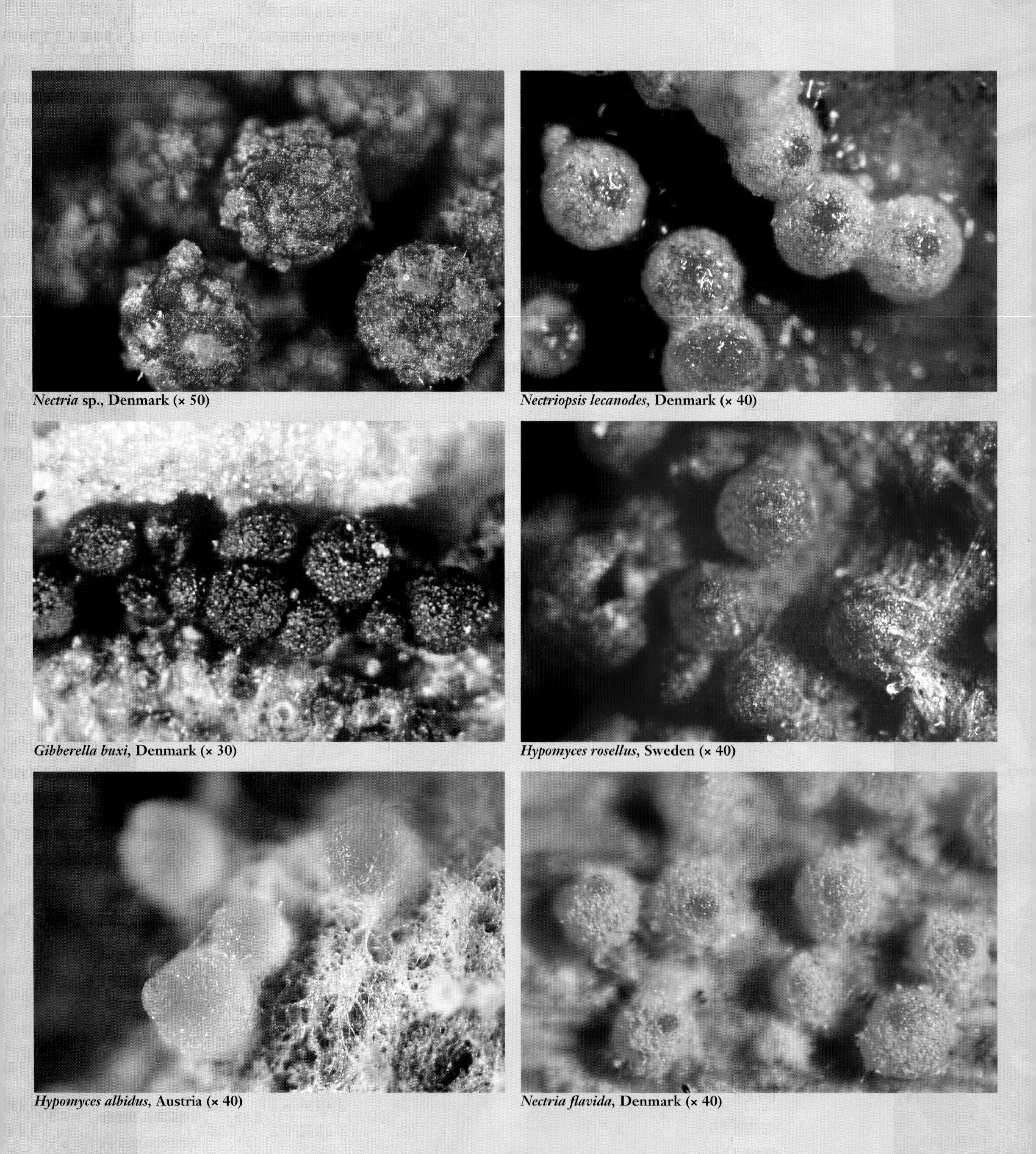

Nectria sp., Denmark (× 50)

Nectriopsis lecanodes, Denmark (× 40)

Gibberella buxi, Denmark (× 30)

Hypomyces rosellus, Sweden (× 40)

Hypomyces albidus, Austria (× 40)

Nectria flavida, Denmark (× 40)

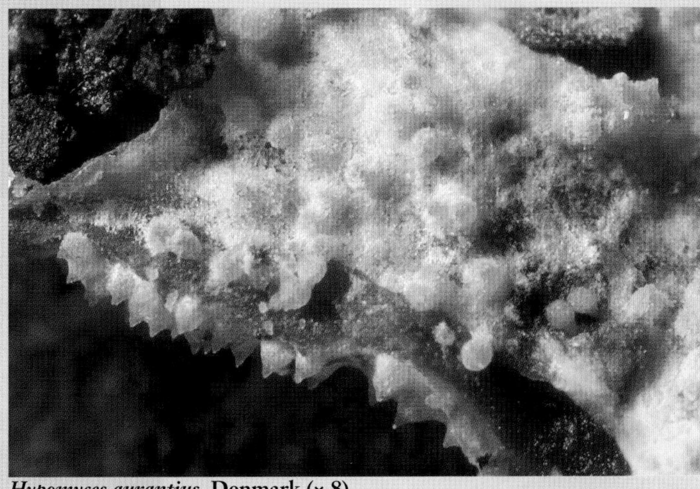

Hypomyces aurantius, Denmark (× 8)

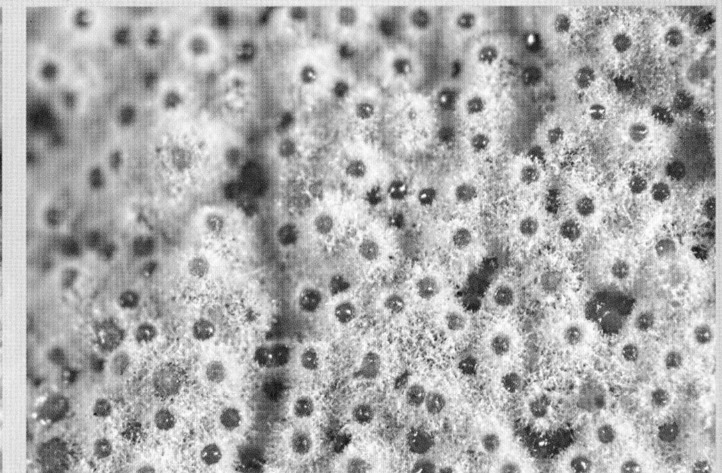

Nectriopsis broomeana, Denmark (× 10)

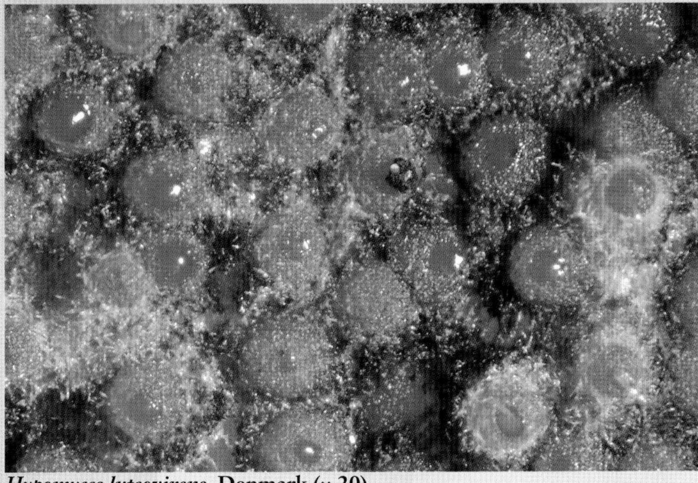

Hypomyces luteovirens, Denmark (× 20)

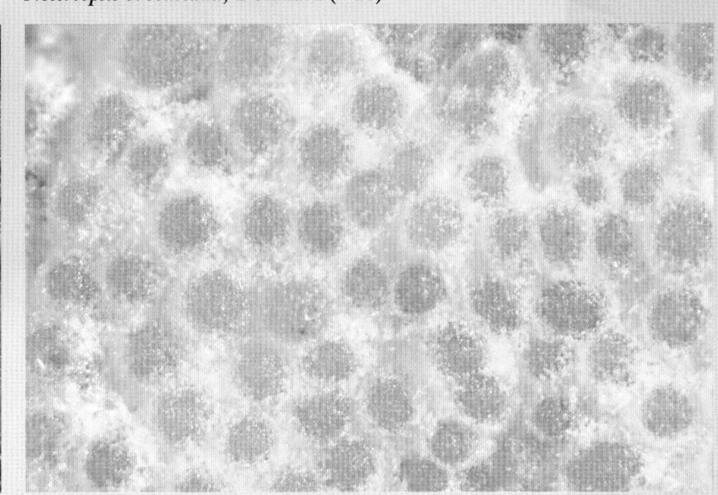

Hypocrea pulvinata, Denmark (× 20)

Nectria sinopica, Denmark (× 40)

Hypocrea spinulosa, Denmark (× 6)

Cordyceps locustiphila on a grasshopper, Ecuador (× 4)

Cordyceps militaris on butterfly larva, Denmark (× 3)

The genus *Cordyceps* and its close allies are one of the more fascinating groups of fungi. They are parasites that attack living hosts, with the majority of species infecting and killing insects. After the fungus kills the host, large, colored stromata arise. Here, the spotted heads are filled with the small, spore-producing fruiting bodies.

This group is relatively scarce in temperate areas, whereas its tropical diversity is immense.

Ophiocordyceps amazonica on a grasshopper, Ecuador (× 8)

Cordyceps sp. on a stick insect, Ecuador (× 1)

Metacordyceps chlamydosporia on insect eggs, Ecuador (× 3)

Ophiocordyceps australis on an ant, Ecuador (× 6)

Ophiocordyceps gracilis on a moth larva, Sweden (× 1.2)

Cleistothecial fungi

The fruiting bodies of cleistothecial fungi are closed spheres. At maturity, these either split in some way to allow the asci to shoot out the spores, or they may simply rot away, allowing the spores to spread without being actively propelled.

The fruiting bodies are very small and will often serve as dispersal agents themselves.

Among the species with active spore dispersal are the powdery mildews. These live as parasites on leaves and stems of living plants. At the end of the year they form cleistothecial fruiting bodies as a resting stage for the winter. In the spring the fruiting bodies swell and crack open so the asci can shoot the spores onto newly emerging plants.

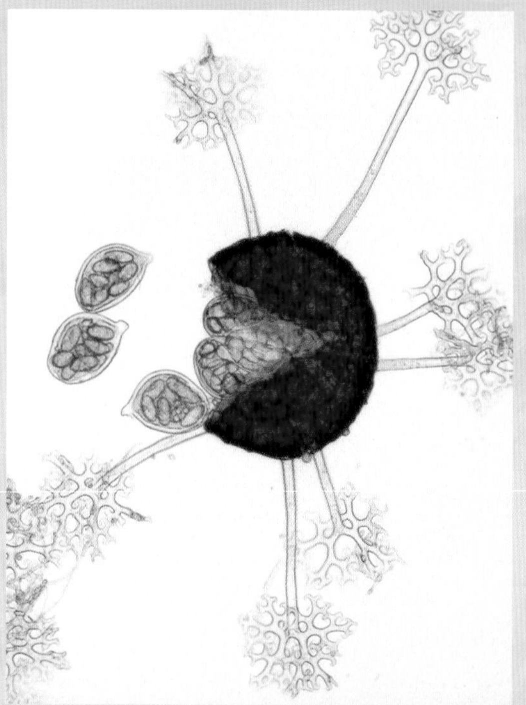

Erysiphe alphitoides with asci containing eight spores (× 50)

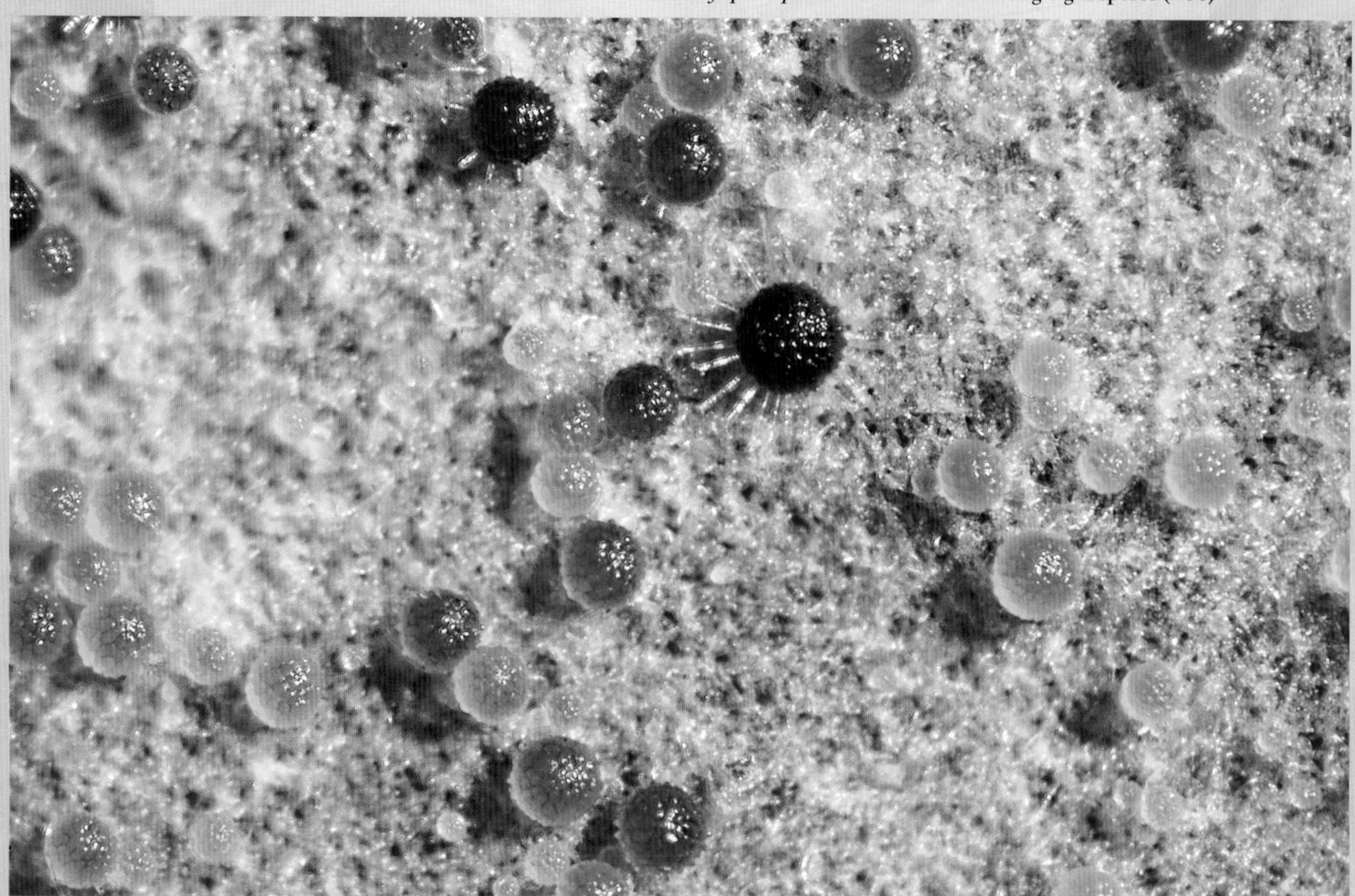

Erysiphe alphitoides, a parasite on *Quercus* leaves, Denmark (× 30)

Phyllactinia guttata on stilts and with a sticky area at the top, Denmark (× 70)

Erysiphe palczewskii on legumes, Denmark (× 80)

Arthroderma curreyi, Denmark (× 40)

Auxarthron sp., Denmark (× 60)

Eurotium cf. *amstelodami*, Denmark (× 50)

Aphanoascus fulvescens, Denmark (× 30)

Gymnoascus reessii, Denmark (× 60)

Talaromyces luteus, Denmark (× 30)

A large group of cleistothecial fungi are specialists in decomposing keratin—an essential compound in skin, nails, and horn (this page). These fungi have totally passive spore dispersal and thus often have small, more or less reduced asci.

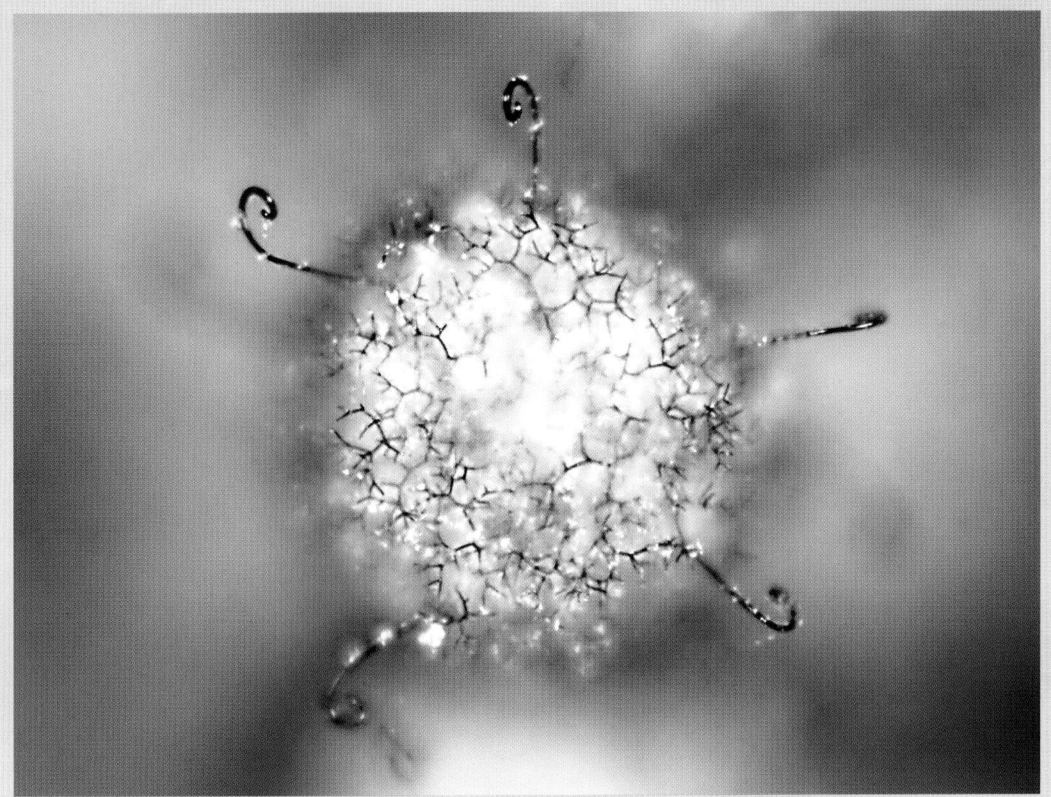

Myxotrichum chartarum, Denmark (× 130)

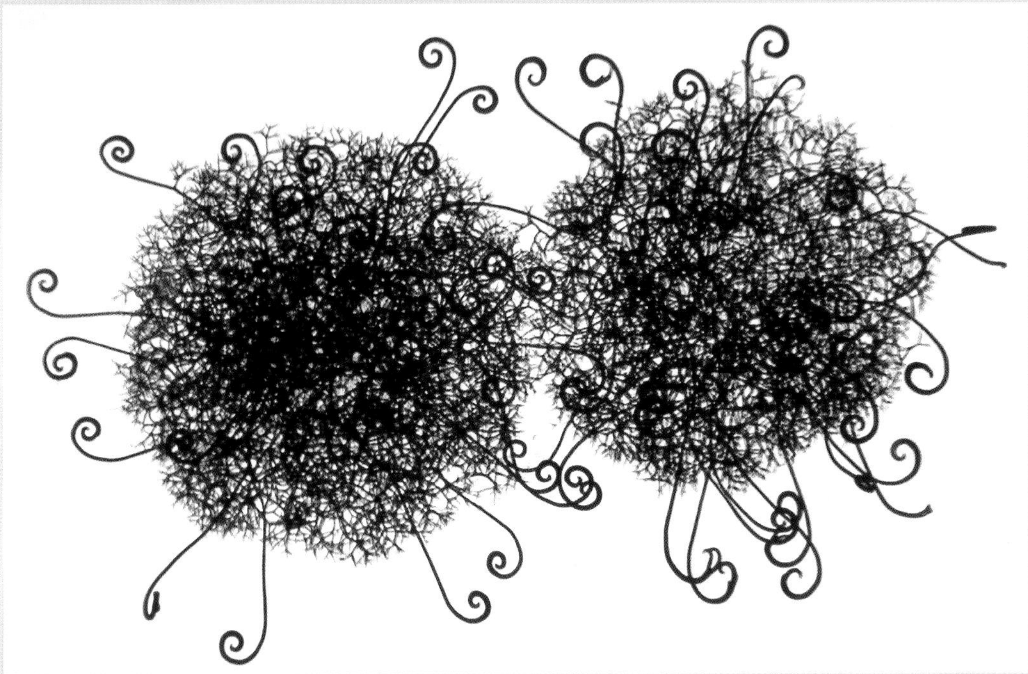

Myxotrichum chartarum seen in the light microscope, Denmark (PBH, × 110)

Common to many cleistothecial fruiting bodies are specialized appendages that enable them to cling to a dispersal agent, such as insects or birds.

Discolichens

Lichens are weird double organisms consisting of a fungus and a green alga (or more rarely a cyanobacterium, or both). The fungus builds the tissue (*thallus*) that houses the photobionts. The lichens are the only type of fungus that produces large, fleshy structures not connected to reproduction.

Lichens may also form fruiting bodies. As most lichens are related to the cup fungi, they often form disk-shaped fruiting bodies, typically in the centers of old thalli. Lichens with disk-shaped fruiting bodies are called discolichens.

Fruiting bodies of this type will produce spores in asci within a palisade tissue at the surface of the disk. The fruiting body may contain a layer of algae below the hymenium. See more on lichens on page 218.

Thallus of *Melanelia fuliginosa*, Denmark (× 3)

Thallus of *Peltigera praetextata*, Sweden (× 1)

Mixed alpine lichens, some with dark, disk-shaped fruiting bodies, Austria (× 1)

Dibaeis cf. *columbiana* with gelatinous pink fruiting bodies, montane Ecuador (× 2)

Although lichens may tolerate very harsh and dry habitats (the extreme being Arctic deserts), they thrive in wet, temperate climates such as the cold coniferous zones of North America, Europe, and Asia or the mountain forests of the tropics.

Previous spread : *Cladonia* sp., with red fruiting bodies at the tips, Ecuador (× 4)

Thalli of *Usnea* sp., montane Ecuador

Mazaedioid fungi

The mazaedioid fungi are Ascomycota where the asci decay, leaving the powdery spores on top of the fruiting body. The fruiting bodies are mostly very small and resemble miniature pins.

Many mazaedioid species are drought resistant. Some are lichenized (these pages), while others decompose various dead substrata, for example, sun-exposed wood or resin.

Chaenotheca sp., Sweden (× 50)

Calicium viride, Denmark (× 70)

Calicium salicinum, Denmark (× 30)

Calicium sp., Sweden (× 70)

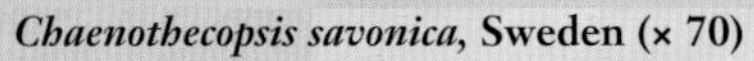

Chaenothecopsis savonica, Sweden (× 70)

Mycocalicium subtile, Sweden (× 70)

Chaenothecopsis debilis, Sweden (× 70)

Chaenothecopsis caespitosa, England (× 10 and 30)

The mazaedioid fungi on these pages are decomposers and are not closely related to the lichenized species.

Truffles

Truffles form tuber-shaped, underground fruiting bodies without active spore release. They are dispersed by animals (e.g., rodents, pigs, and deer), which dig them up and eat them after being attracted by the strong odors emitted by the mature fruiting bodies.

Truffles are fungi that, in the course of evolution, have kept the production of their fruiting bodies belowground. This has happened many times in fungal evolution (parallel evolution, see page 40), and there are numerous truffle species within the Ascomycota, the Basidiomycota, and the Zygomycota (the latter two are often called false truffles). Molecular phylogenies show that many truffles belong within traditional genera like *Cortinarius* or *Peziza*, between "normal" species with active spore release. Almost all truffles form ectomycorrhiza (see page 210).

Truffles that belong to the ascomycote genus *Tuber* are among the most valued fungi in gastronomy. Trained dogs are used to find truffles by their smell, as are pigs, though these were used more frequently in the past.

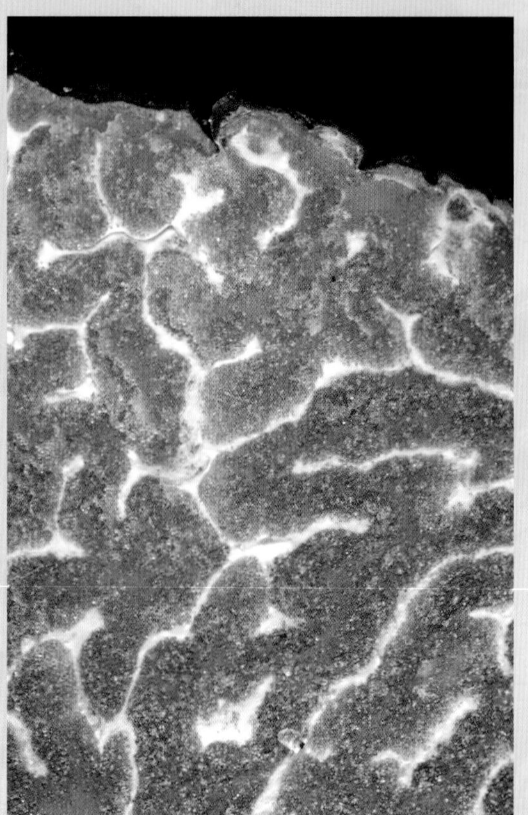

Tuber aestivum, Denmark (× 10)

Tuber aestivum, Denmark (JV, × 1.5)

Choiromyces venosus, Denmark

Species of the genus *Elaphomyces*—deer truffles—are the easiest truffles to find in nature. Just look for the parasitic *Elaphocordyceps* species sticking out of the ground, dig down five to ten centimeters, and you have the truffle! Unfortunately, deer truffles are not edible.

Elaphomyces granulatus with the parasitic fungus *Elaphocordyceps ophioglossoides*, Denmark

Laboulbeniomycetes

The last group within the Ascomycota is a peculiar group with a peculiar name: the Laboulbeniomycetes. They are tiny, about half a millimeter, and grow on living insects—typically beetles, more rarely flies, grasshoppers, termites, cockroaches, or even other arthropods like mites or millipedes. They are placed in their own class within the Ascomycota.

A laboulbeniomycete forms a few thick, branched hyphae (*haustoria*), which may penetrate the exoskeleton of the host animal, and a small fruiting body that contains the cells necessary for sexual reproduction, including the asci where the spores are formed. Even though they have been shown to incorporate small amounts of nutrients from their host, they appear to do no real harm.

There are approximately two thousand described species of Laboulbeniomycetes. Many of these are specific to certain host genera or even to specific species. Assuming there are two million beetles on the planet and one out of fifty has a specialized fungal symbiont, the expected number of Laboulbeniomycetes should be around forty thousand. Thus many of the fungi that are yet to be discovered might hide in this strange class (more on the expected number of species on page 254).

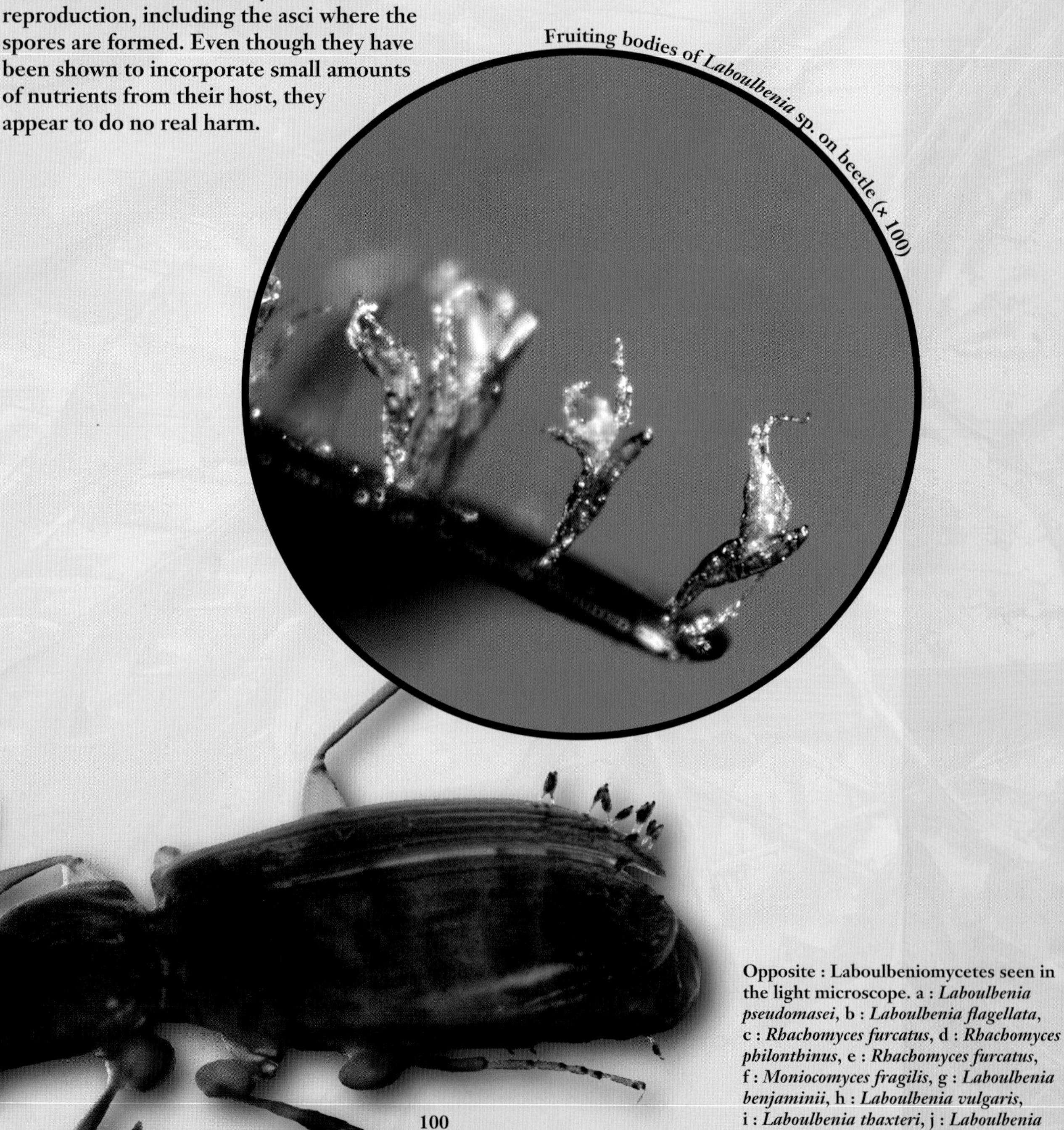

Fruiting bodies of *Laboulbenia* sp. on beetle (× 100)

Opposite : Laboulbeniomycetes seen in the light microscope. a : *Laboulbenia pseudomasei*, b : *Laboulbenia flagellata*, c : *Rhachomyces furcatus*, d : *Rhachomyces philonthinus*, e : *Rhachomyces furcatus*, f : *Moniocomyces fragilis*, g : *Laboulbenia benjaminii*, h : *Laboulbenia vulgaris*, i : *Laboulbenia thaxteri*, j : *Laboulbenia leisti*, Denmark (JH, × 100–200)

a
b
c
d
e
f
g
h
i
j

The Basidiomycota

The Basidiomycota are defined by the production of sexual spores on *basidia*. Basidia are mostly club-shaped cells with four small outgrowths called *sterigmata*. The sterigmata produce one spore each. The spores are discharged actively but with much less force than in the Ascomycota.

Another feature unique to the Basidiomycota is the presence of *clamp connections* in many species. Clamps are small outgrowths at the transverse separations (*septa*) of the hyphae. They help the nuclei behave correctly during cell divisions.

The Basidiomycota form the second largest group of fungi, with more than thirty-one thousand described species. Boletes, agarics, polypores, chantarelles and puffballs are good examples of the Basidiomycota.

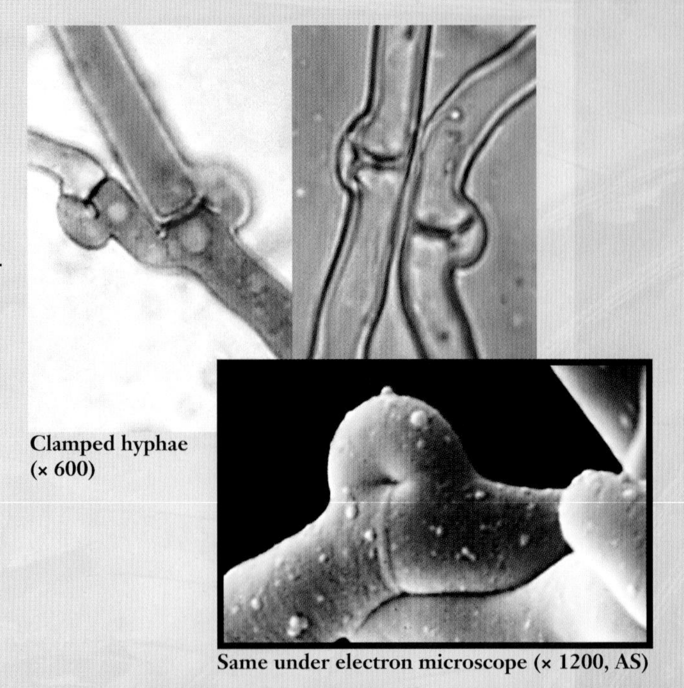

Clamped hyphae (× 600)

Same under electron microscope (× 1200, AS)

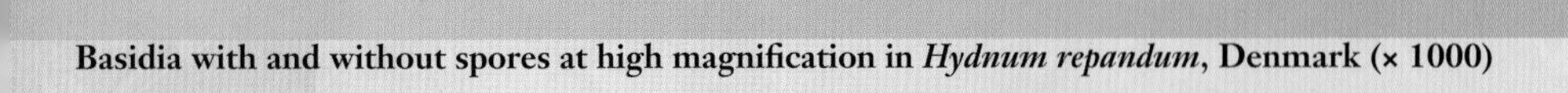

Basidia with and without spores at high magnification in *Hydnum repandum*, Denmark (× 1000)

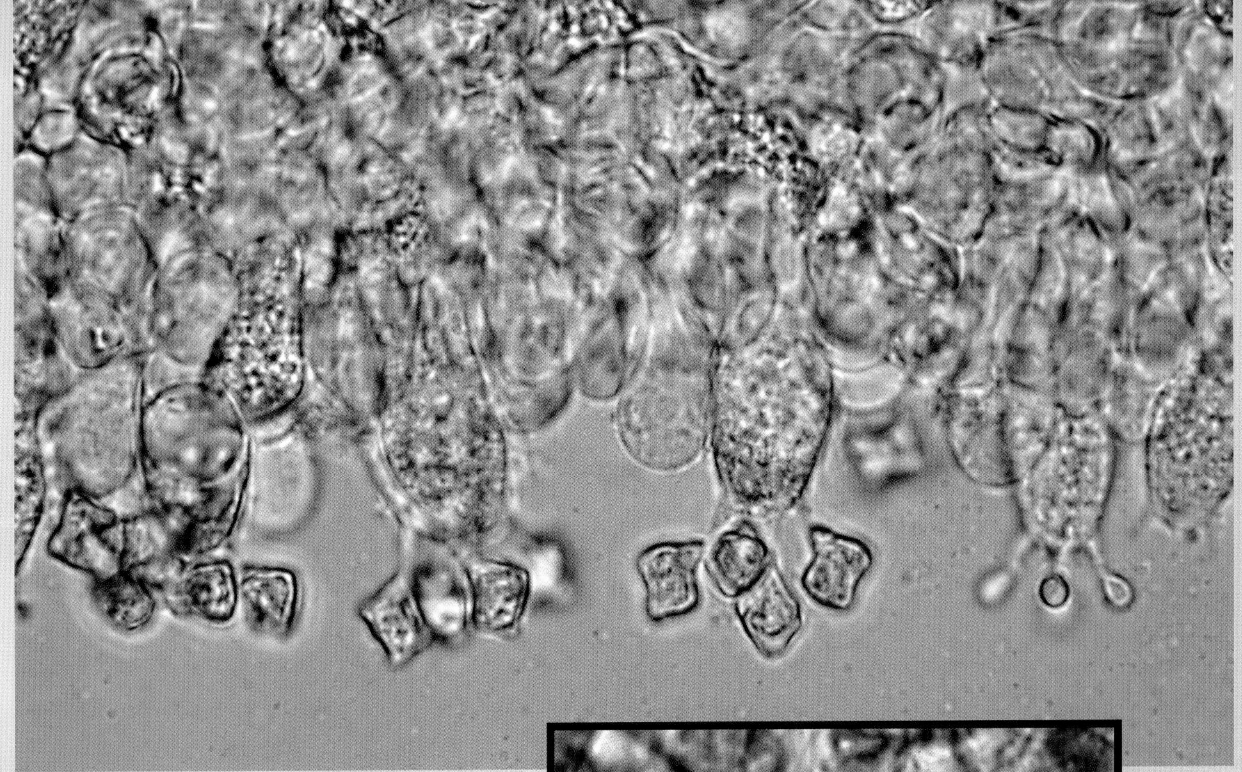

Hymenium with basidia bearing the angular spores of *Entoloma conferendum*, Denmark (× 1000)

The basidia are normally found in a hymenium, where they sit parallel to each other intermixed with sterile, club-shaped *basidioles*. Because spore release is rather weak, the hymenium is usually found on a vertical or downward-pointing surface so the spores can fall freely after release.
It also helps if the hymenium is placed high, for example under a cap on top of a stem.

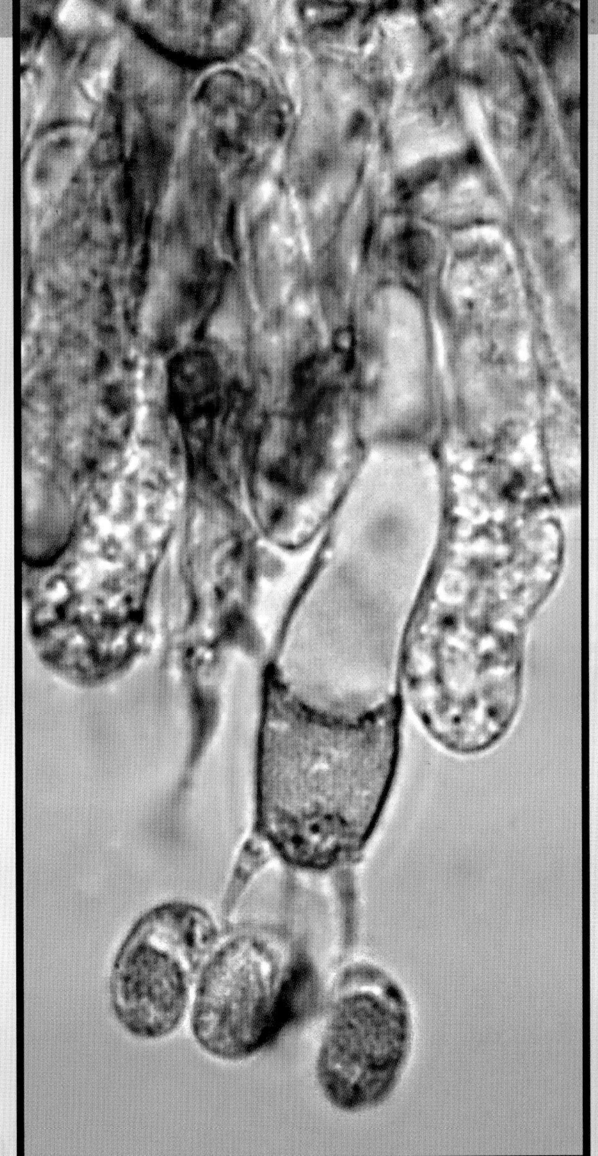

4-spored basidium, *Hygrocybe conica*, Denmark (× 2500)

The basidiomycote hymenium may be configured in different ways. The simplest version is just a smooth surface—easy to build but with very little surface area. This type of hymenium is found in various corticioid, or "crust," fungi.

Many species will, however, try to optimize spore production by folding the hymenium, thus making a larger surface. This can be done by placing the hymenium on spines or gills or in tubes or pores. If you study a fresh gill under high magnification you can actually see the spores sitting in groups of four on the basidia.

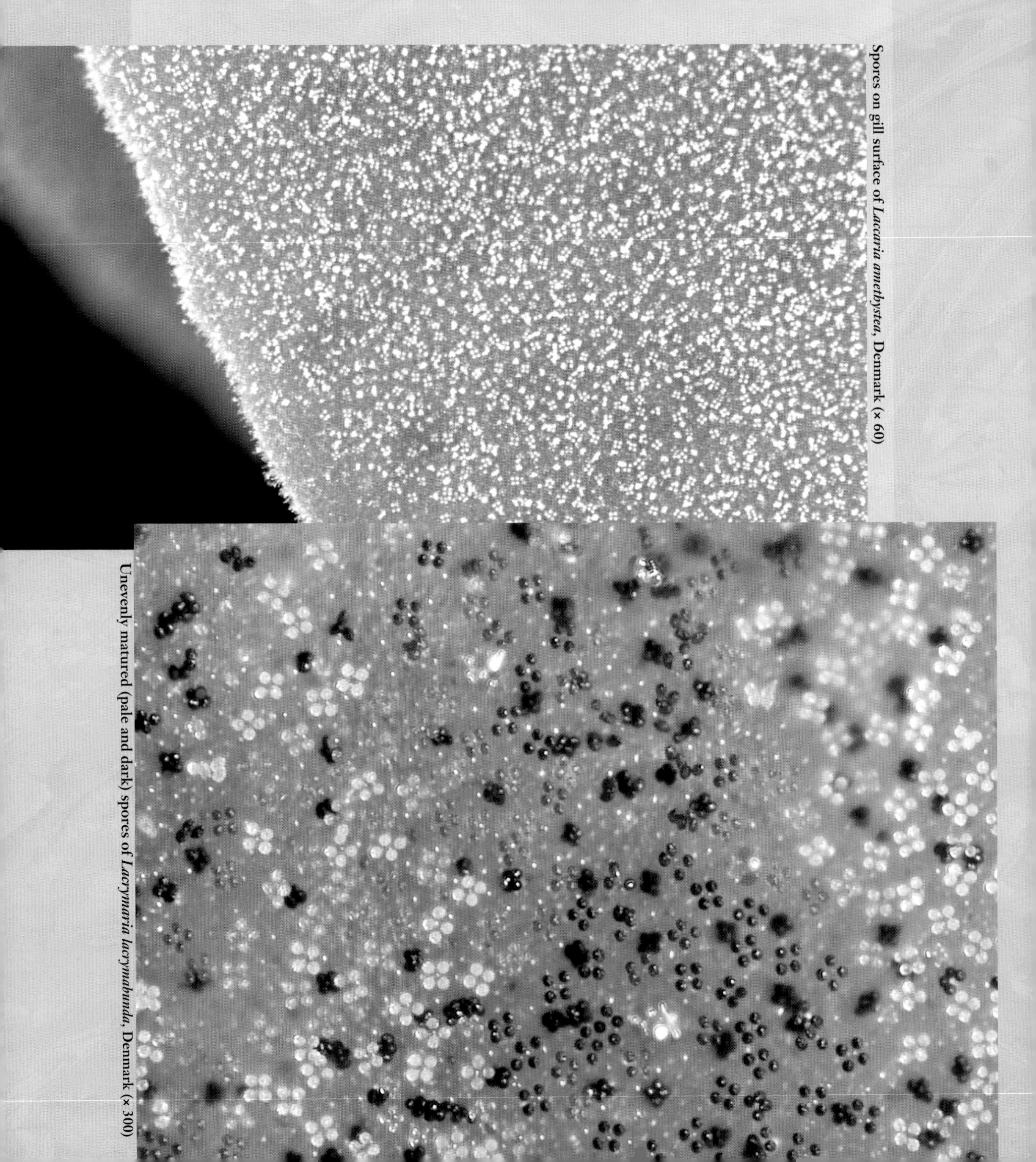

Spores on gill surface of *Laccaria amethystea*, Denmark (× 60)

Unevenly matured (pale and dark) spores of *Lacrymaria lacrymabunda*, Denmark (× 300)

The Basidiomycota are divided into three classes: the rusts (Pucciniomycotina), the smuts (Ustilagomycotina), and the rest (Agaricomycotina). The smuts and rusts are generally plant parasites and rarely form fruiting bodies. All fungi with large, fleshy fruiting bodies that we normally recognize as Basidiomycota belong to the Agaricomycotina.

The Agaricomycotina is again divided into three groups, which are recognized by their different types of basidia: the Tremellomycetes, the Dacrymycetes, and the Agaricomycetes. The well known agarics, chanterelles, boletes, polypores, and club fungi all belong in the Agaricomycetes.

On the following pages the fruiting body types of the Basidiomycota are illustrated.

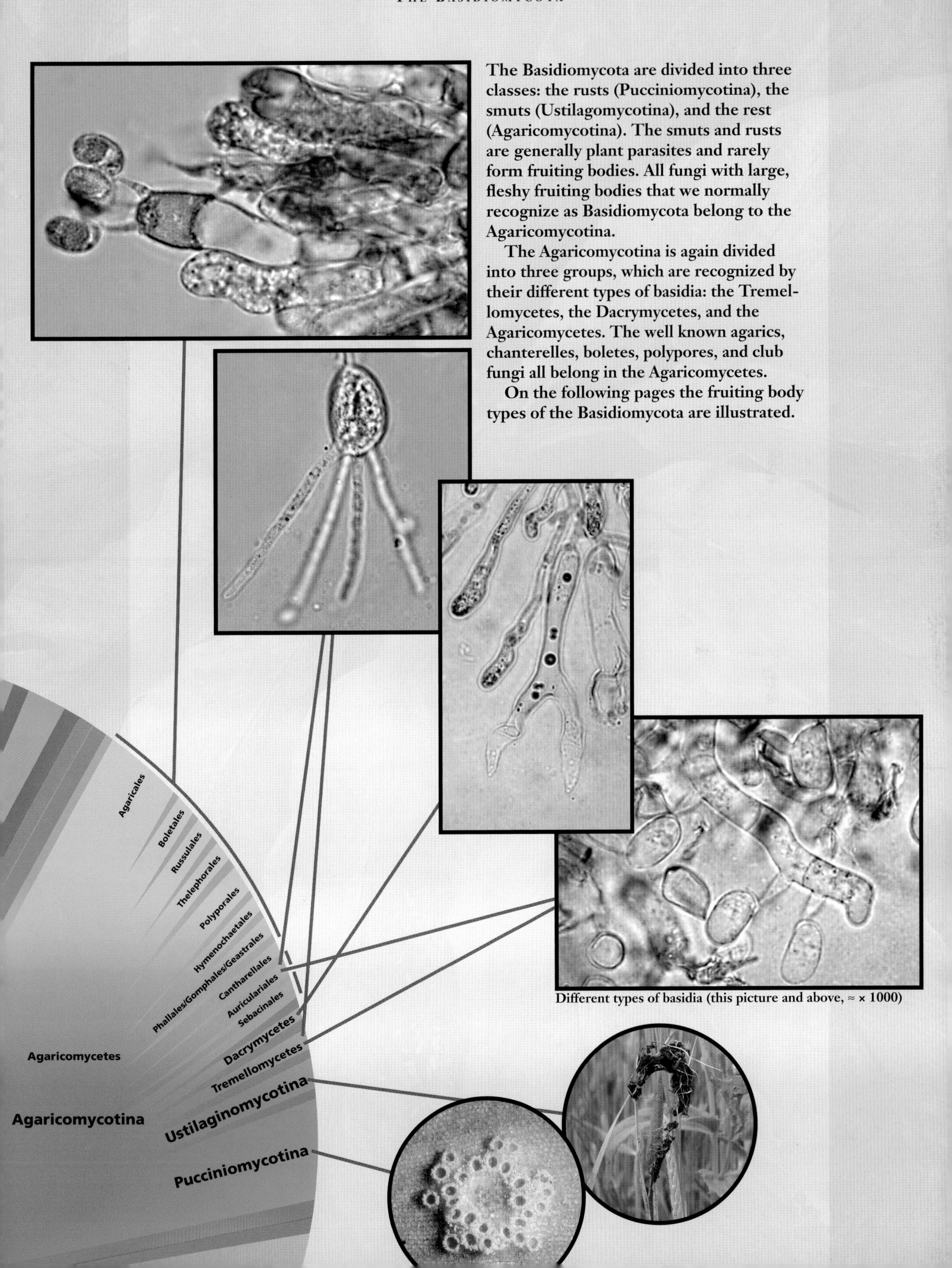

Different types of basidia (this picture and above, ≈ × 1000)

Corticioid fungi

The corticioid fungi (also called crust fungi or resupinate fungi) form flat fruiting bodies. These are found mostly on the underside of trunks and branches, which keep them off the ground.

Hyphodontia sambuci, Denmark (× 1)

Peniophora cinerea, Denmark (× 1)

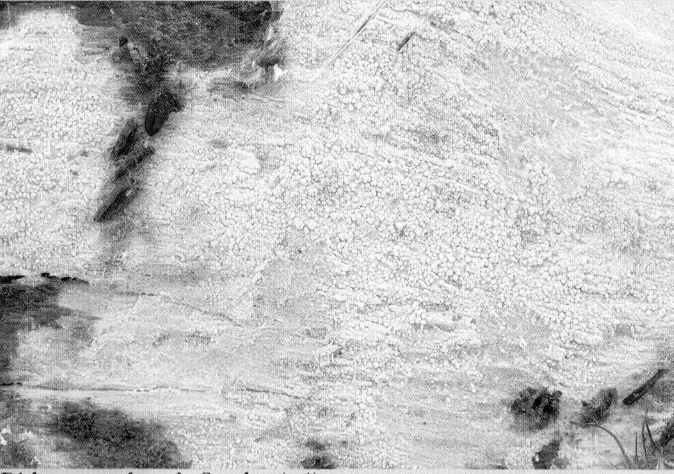

Dichostereum boreale, Sweden (× 1)

Amaurodon mustialaensis, Denmark (TL, × 1)

Peniophora incarnata, Denmark (× 1)

Phlebia subochracea, Denmark (× 2)

The surface of the fruiting body is covered by the hymenium, which may be placed on a smooth, wrinkled, warty, or spiny layer—the more complex, the greater the surface area.

Peniophora polygonia, Denmark (× 2)

Tomentella lateritia, Denmark (× 15)

Resinicium bicolor, Denmark (× 5)

Epithele typhae, Denmark (× 15)

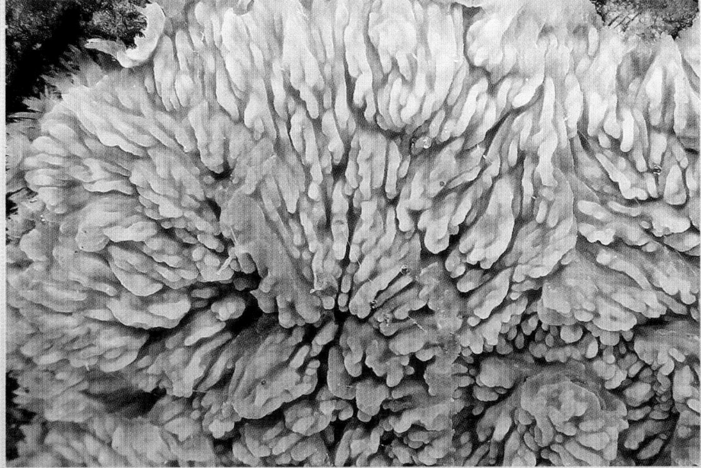
Phlebia radiata, Denmark (× 5)

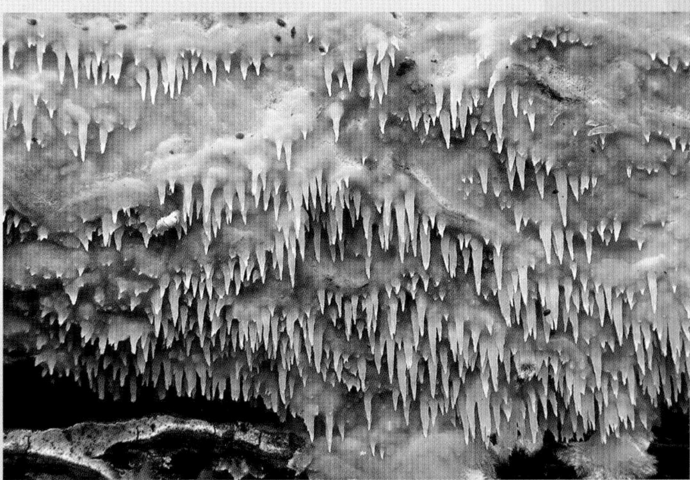
Phlebia uda, Denmark (× 10)

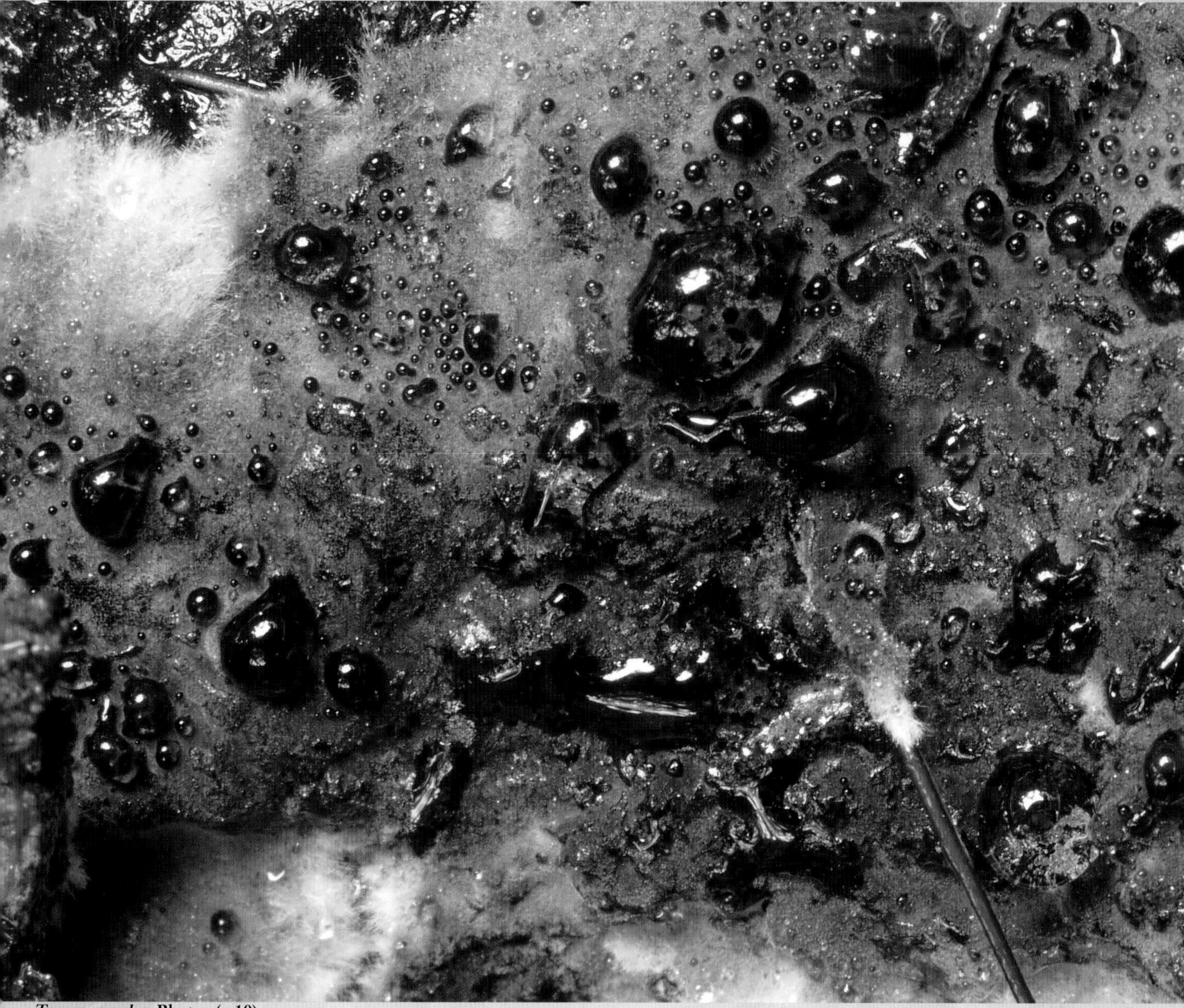

Terana caerulea, Bhutan (× 10)

Tomentella badia, Denmark (× 70)

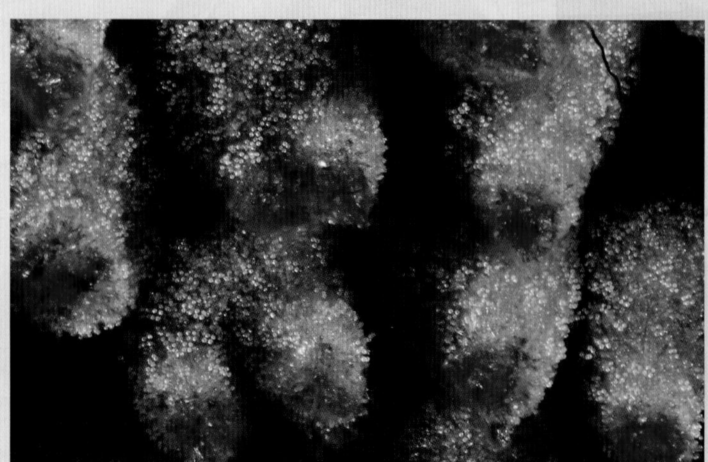

Tomentella crinalis, Denmark (× 20)

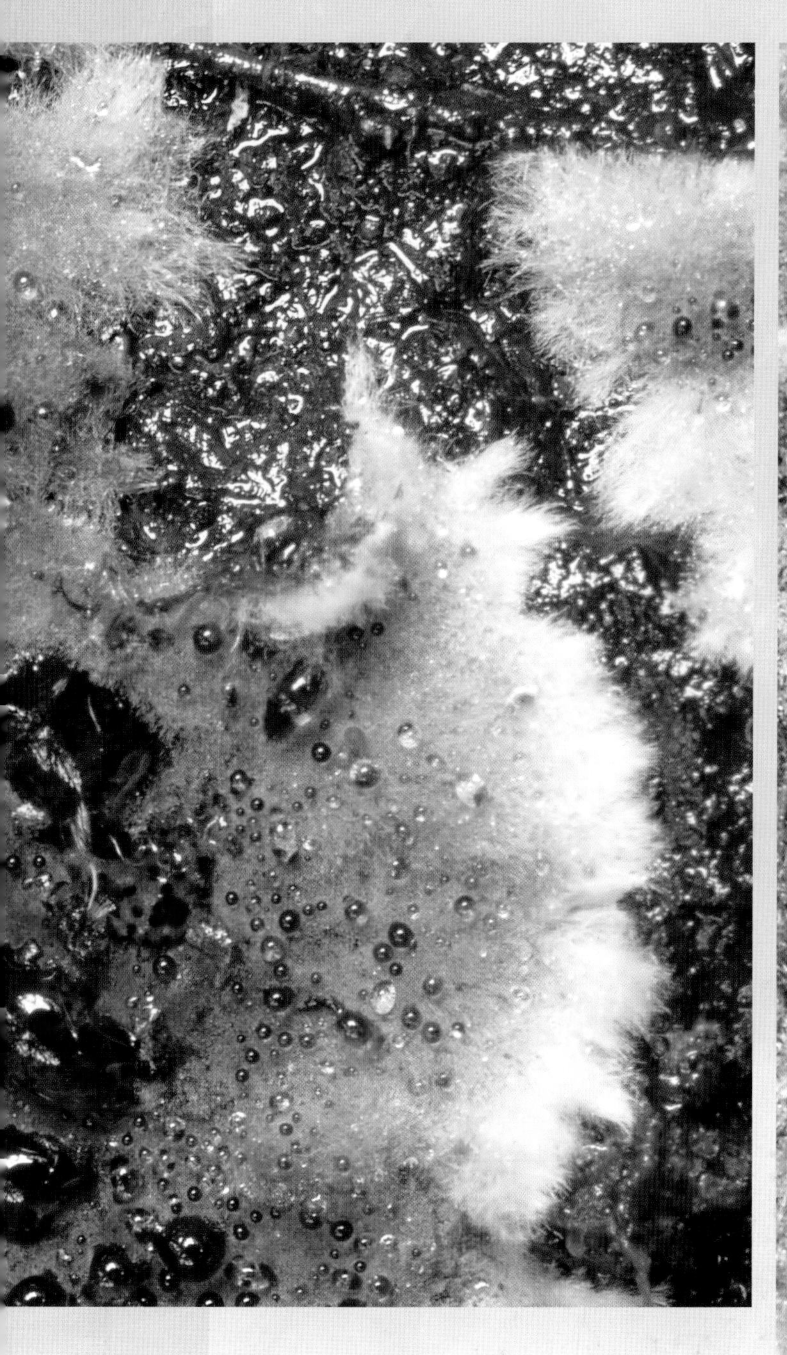

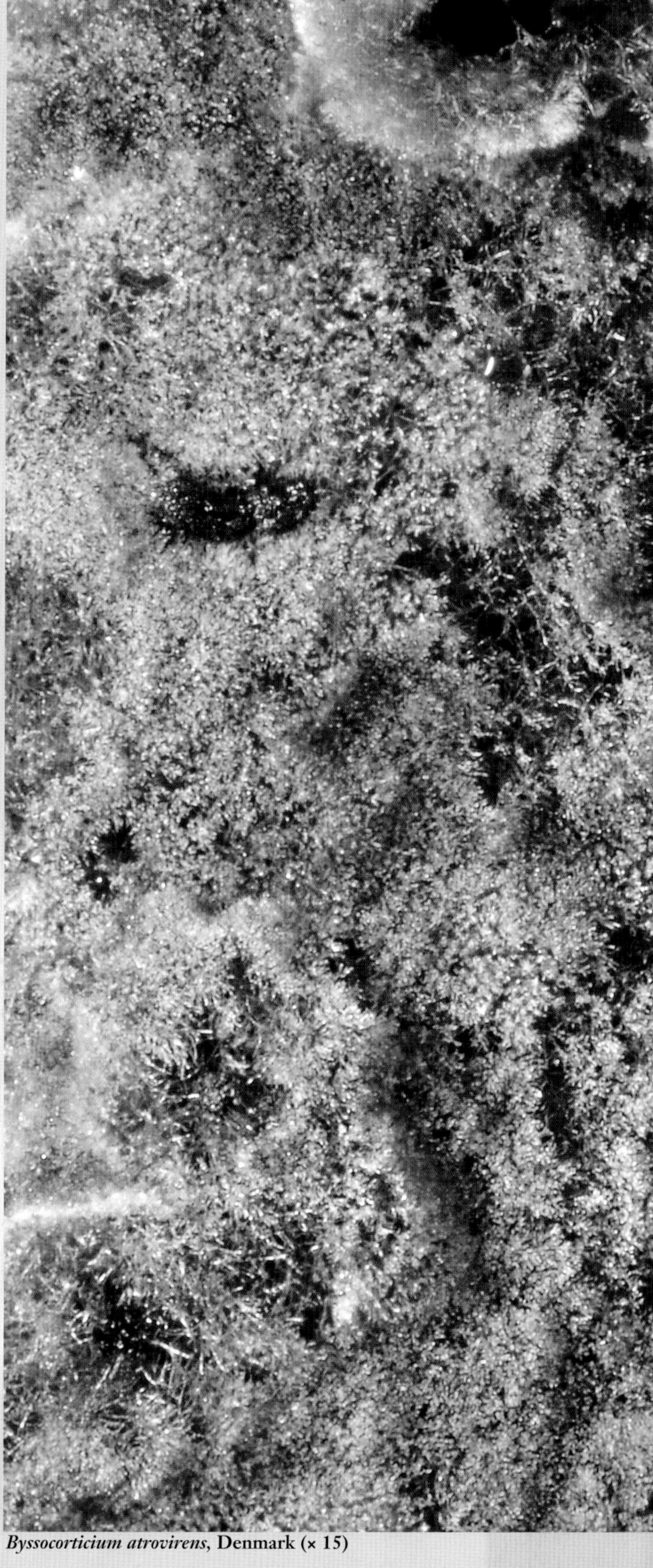

Byssocorticium atrovirens, Denmark (× 15)

The corticioid fungi produce their spores directly on the surface of the fruiting body. At high magnification may be possible to see the spores sitting in groups of four on the hymenial surface (bottom left).

Hymenochaete rubigino
Denmark (×

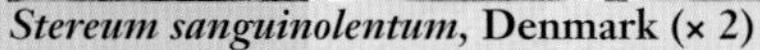
Stereum sanguinolentum, Denmark (× 2)

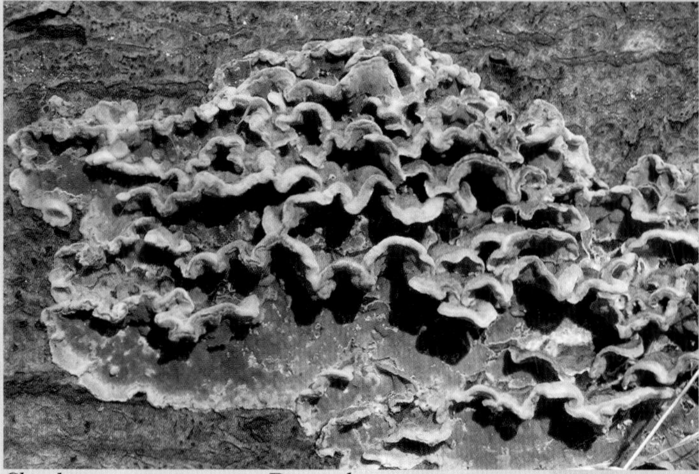
Chondrostereum purpureum, Denmark

Stereoid fungi

The stereoid fungi are an advanced deviation of the corticioids. By forming a small cap at the top of the fruiting body, the stereoid fungi protect the hymenium from rain and at the same time gain a larger spore-producing area. Furthermore, these fungi are often tough and perennial.

Hymenochaete sp. seen from below, Ecuador

Hymenochaete damicornis, Ecuador (× 3)

Cymatoderma dendriticum, Ecuador (× 1)

Tropical species tend to develop even further: tall, stipitate fruiting bodies, sometimes with a basal disk, or even complex, compound fruiting bodies, are quite common.

Some of these are shaped like a small goblet, catching and holding rain to retain moisture longer than would otherwise be possible.

Cotylidia spectabilis, Ecuador (× 2)

Podoscypha nitidula, Ecuador

Stereopsis nigripes, Ecuador (× 4)

Polypores

In polypores, the hymenium is found on the inner side of pores underneath the fruiting body. The fruiting bodies themselves display huge variation: from flat (resupinate) or oblique caps to large, fasciculate or stipitate structures.

Not only do polypores produce some of the largest fruiting bodies, they also make the most long-lived ones. Again the diversity is immense, from the soft-fleshed *Tyromyces* and *Postia* to the perennial, wood-hard *Phellinus*, which can last half a century.

Almost all polypores decompose wood. Some act as parasites that attack and kill living trees, but most decompose trunks and branches that are already dead.

Fomitopsis pinicola, Denmark (× 2)

Trichaptum sp., Bhutan (× 2)

Laetiporus sulphureus, England (× 1)

The pores may vary from tiny—up to ten per millimeter—to two to three millimeters wide. They may be round, edged, labyrinthine, or oblong. In more extreme cases they split with age and form gill-like or irregular toothlike structures.

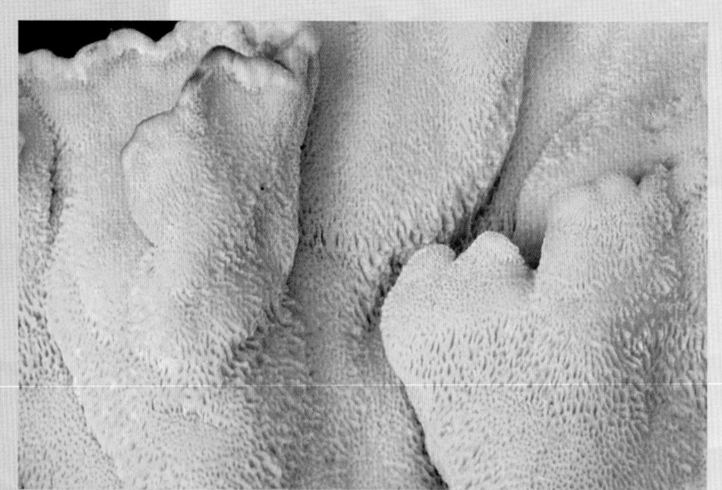

Laetiporus sulphureus, Denmark

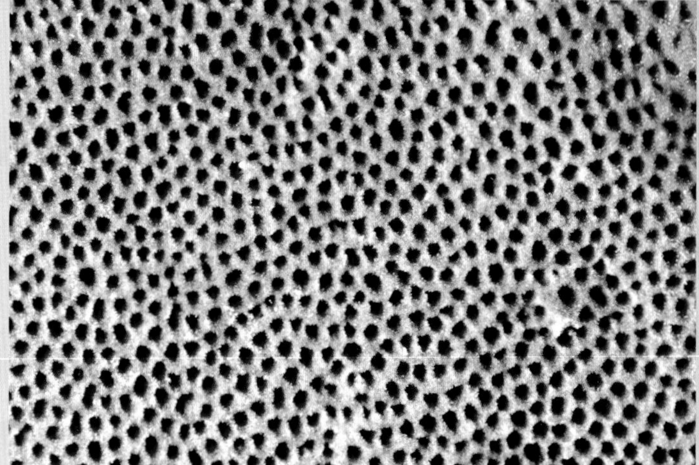

Phellinus pomaceus, Denmark (× 15)

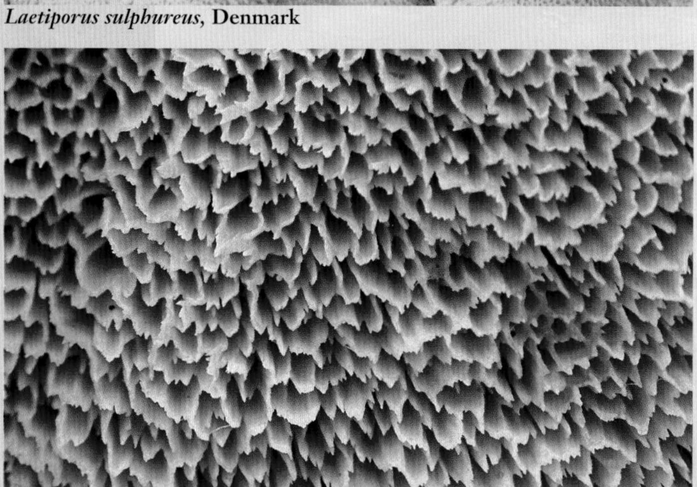

Pycnoporellus fulgens, Denmark (× 8)

Physisporinus vitreus, Denmark (× 3)

Trichaptum abietinum, Denmark (× 8)

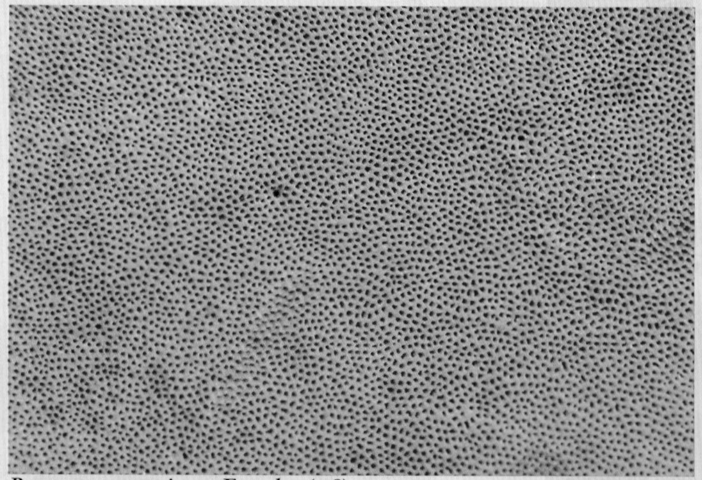

Pycnoporus sanguineus, Ecuador (× 8)

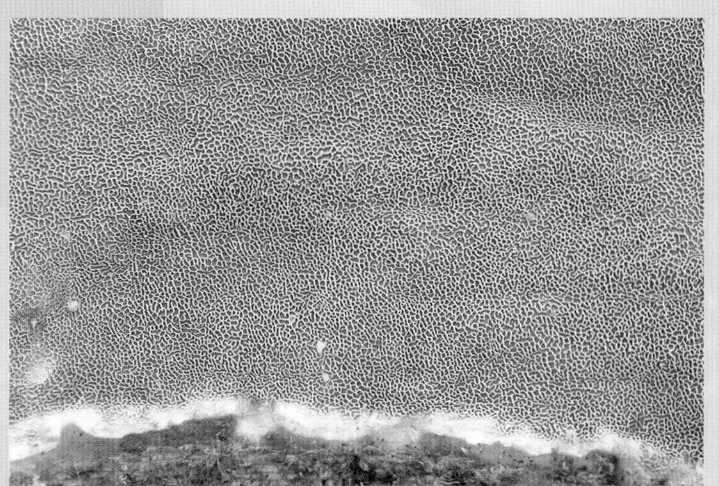
Grammothele fuligo, Ecuador (× 8)

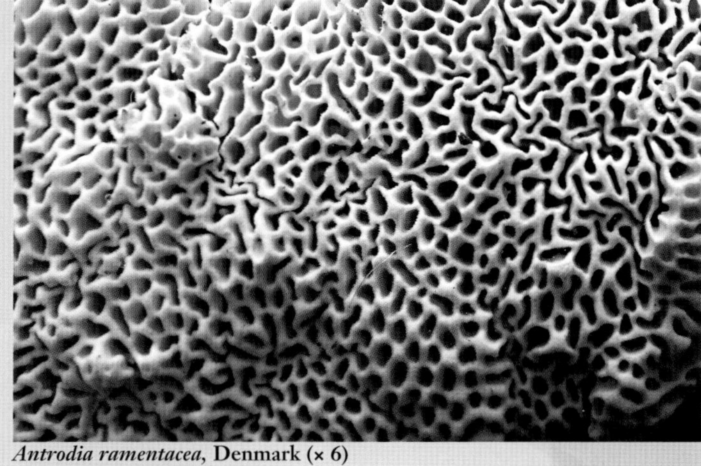
Antrodia ramentacea, Denmark (× 6)

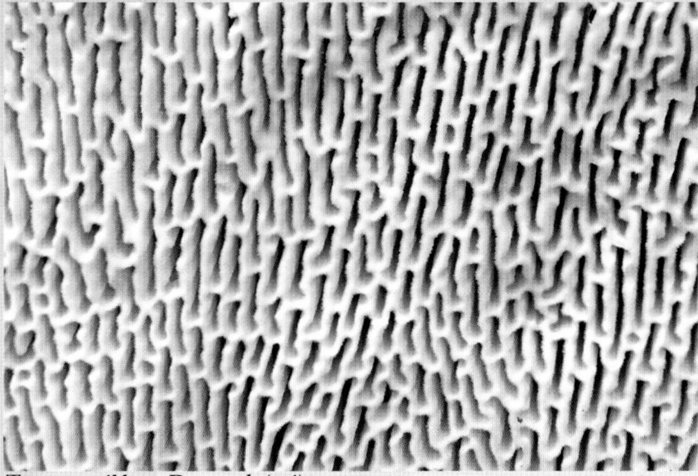
Trametes gibbosa, Denmark (× 4)

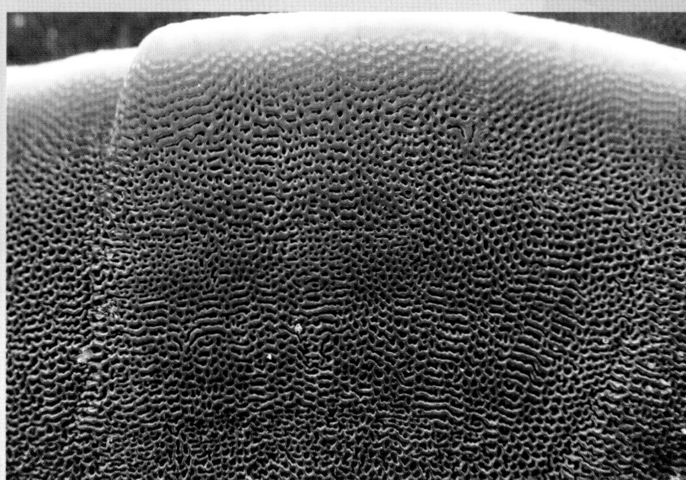
Cyclomyces fuscus, Ecuador (× 8)

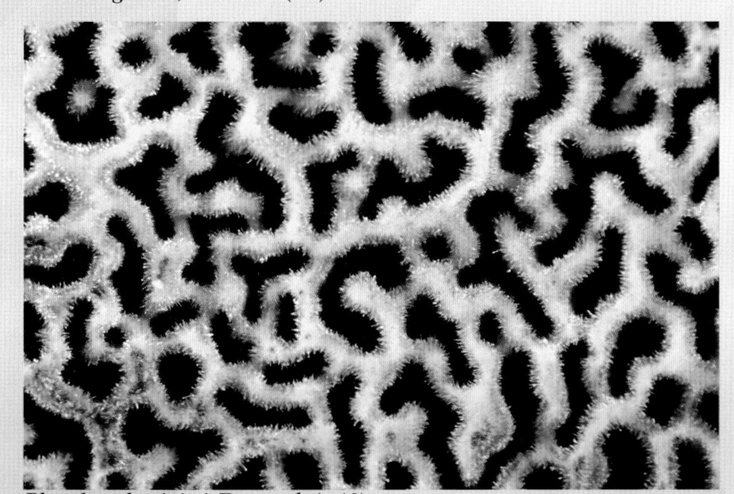
Phaeolus schweinitzi, Denmark (× 12)

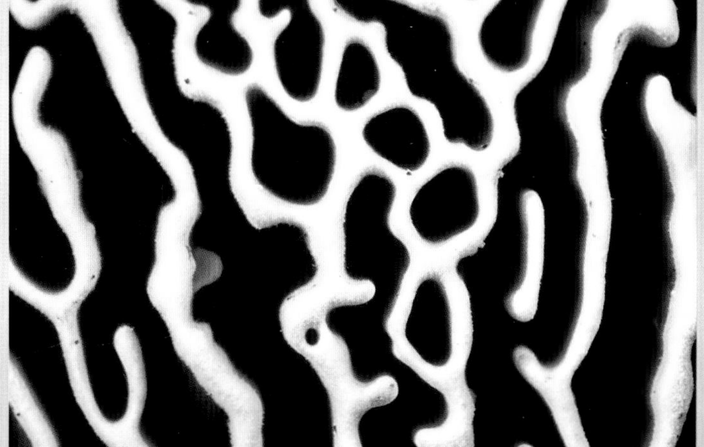
Daedalea quercina, Denmark (× 11)

Grifola sp., Bhutan

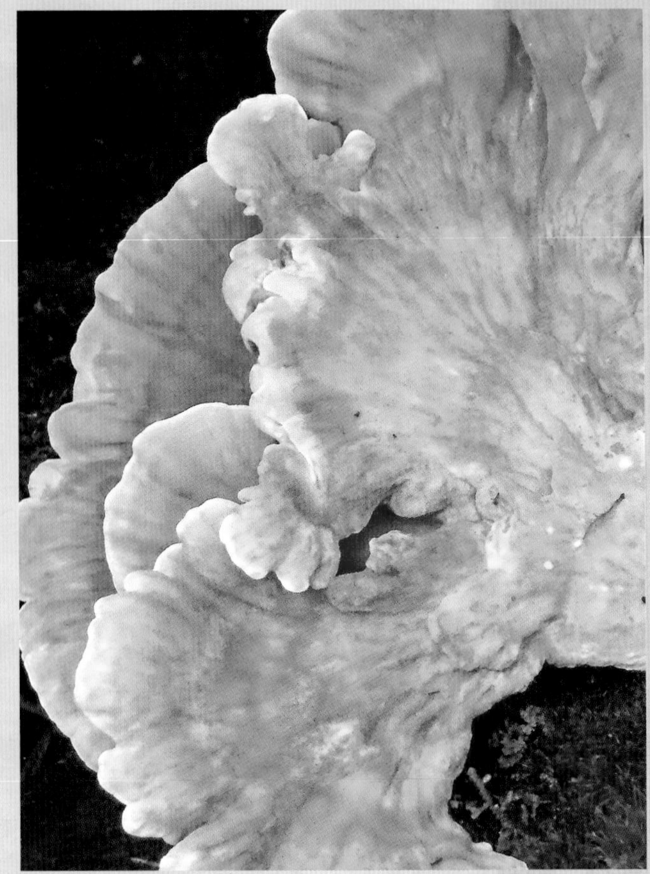

Laetiporus miniatus, Bhutan

The polypores form some of the most beautiful shapes among the fungi ...

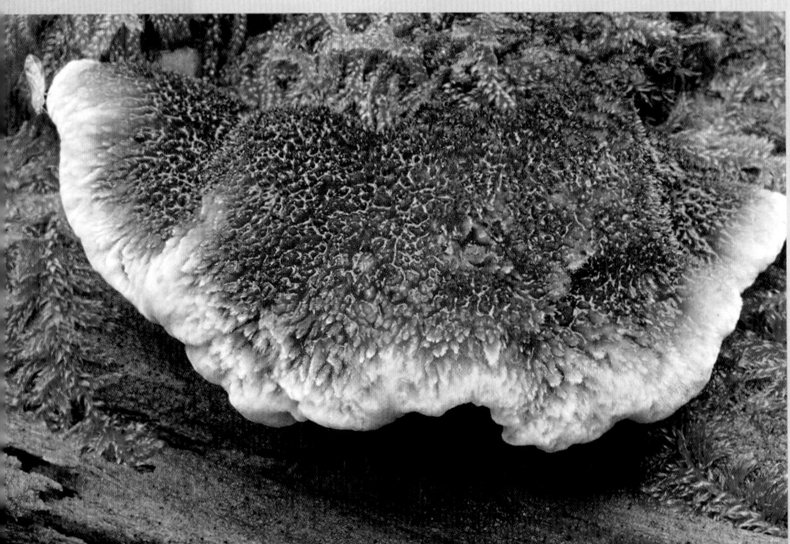

Postia caesia, Denmark (× 2)

Unidentified "Mickey Mouse" polypore, Ecuador (× 2)

Fomes fomentarius, Denmark

Piptoporus betulinus, Denmark

Ganoderma lucidum, Denmark (TL)

Amauroderma sprucei, Ecuador

Ganoderma lucidum coll, Ecuador (× 1.5)

Ganoderma **species are especially eye-catching polypores. They are both elegant and colorful, and often huge. In particular, species in the very diverse** ***Ganoderma lucidum*** **complex, with bright orange stipitate caps covered by a lacquer crust, are among the most beautiful of fungi—and they are found throughout temperate, subtropical, and tropical areas.** ***Ganoderma lucidum*** **and its relatives are even cultivated as medicinal fungi (see page 250).**

Ganoderma adspersum (and the author as a young mycologist), Denmark

Ganoderma pfeifferi, Denmark (TL)

Meripilus giganteus, Denmark

One of the giants among the polypores, *Meripilus giganteus*, regularly creates fruiting bodies best suited for transport by wheelbarrow. And for a time the world's largest fruiting body was claimed to be that of an old *Rigidoporus ulmarius* on an elm stump in Kew Gardens, England. It has since been beaten by another polypore recently found in China.

Meripilus giganteus, Denmark

Rigidoporus ulmarius, Kew Gardens, England

Boletes

The boletes are characterized by their soft fruiting bodies and a layer of tubes beneath the cap. These tubes can be separated from the cap (and often also from each other), thus forming a less rigid structure compared to the tougher polypores.

Most boletes are ectomycorrhizal, needing a mycorrhizal plant partner in order to grow (see page 210). They are generally edible, and several of the most valued edible fungi belong here.

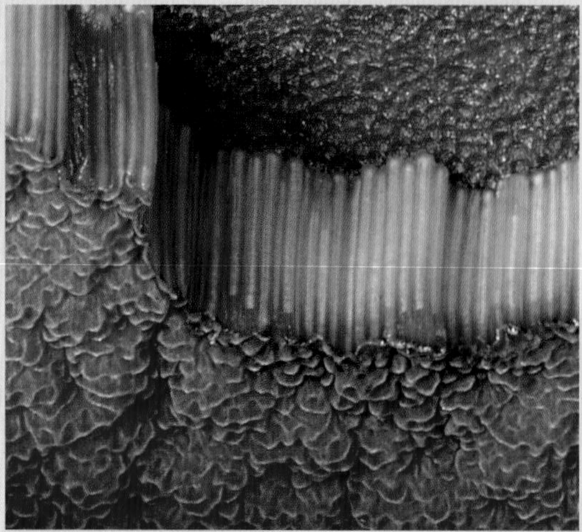
Boletus luridus, Denmark (× 5)

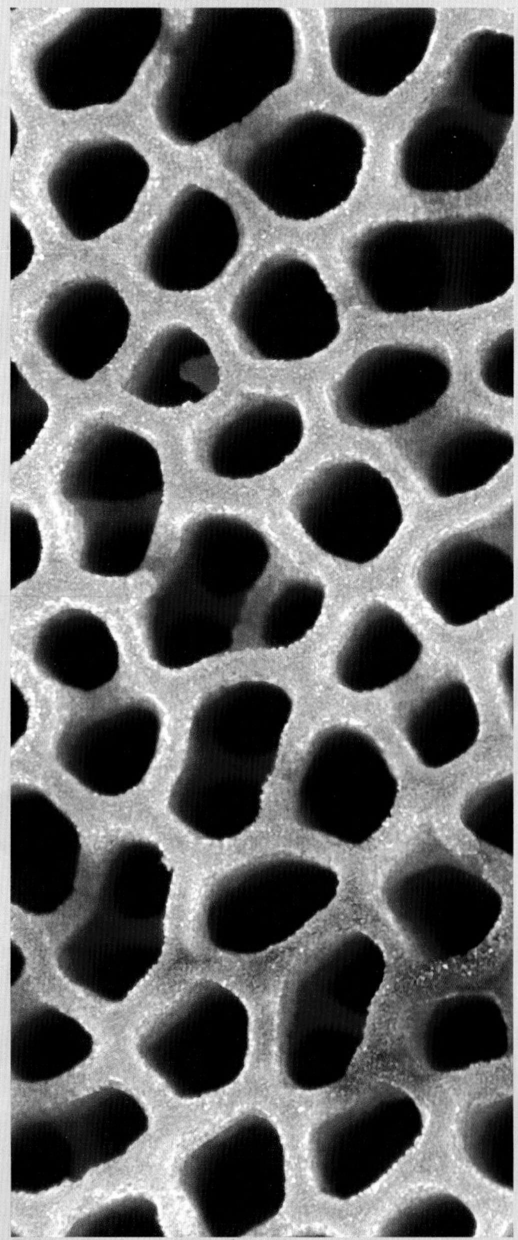
Boletus luridus, Denmark (× 25)

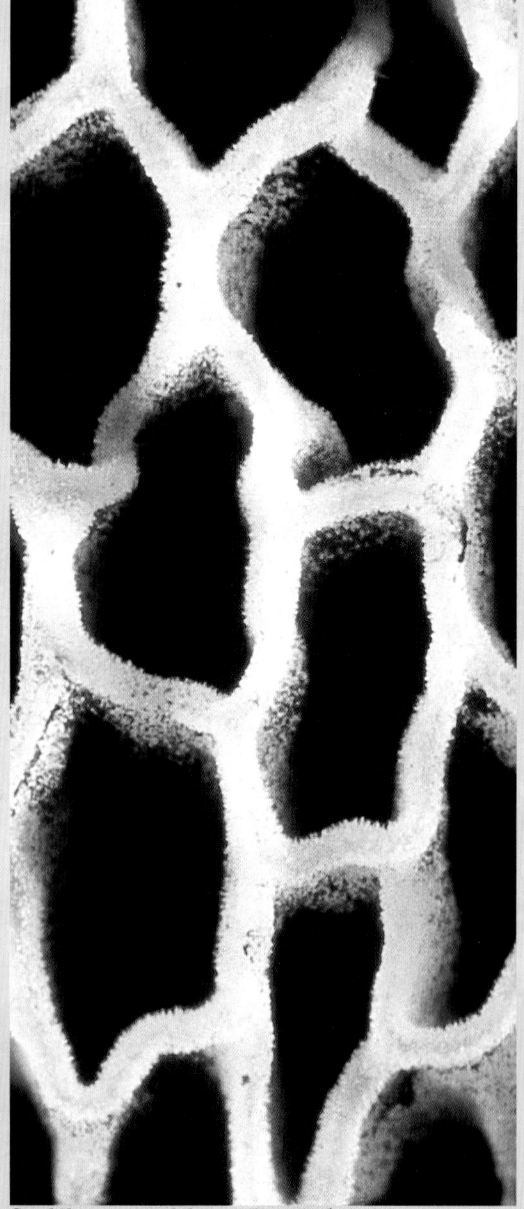
Strobilomyces strobilaceus, Denmark (× 25)

A very good edible species (after cooking): *Boletus luridus*, Denmark

One of the rather few poisonous boletes: *Boletus satanas*, Denmark

Many boletes show strong color reactions when their tissue is wounded. The reactions yield all possible colors but normally fade after several hours. These color changes are very important in the identification of boletes. Contrary to common belief, they are not correlated with edibility: numerous species with strong reactions are well known, edible fungi.

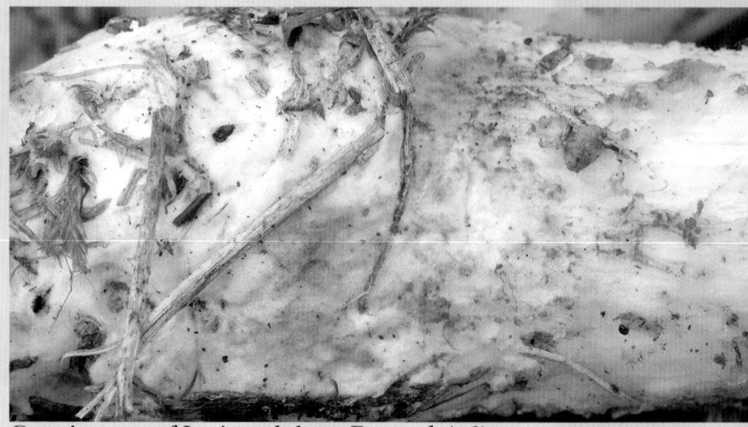

Greening stem of *Leccinum holopus*, Denmark (× 3)

Blueing color variants of *Boletus luridus*, Denmark (× 4)

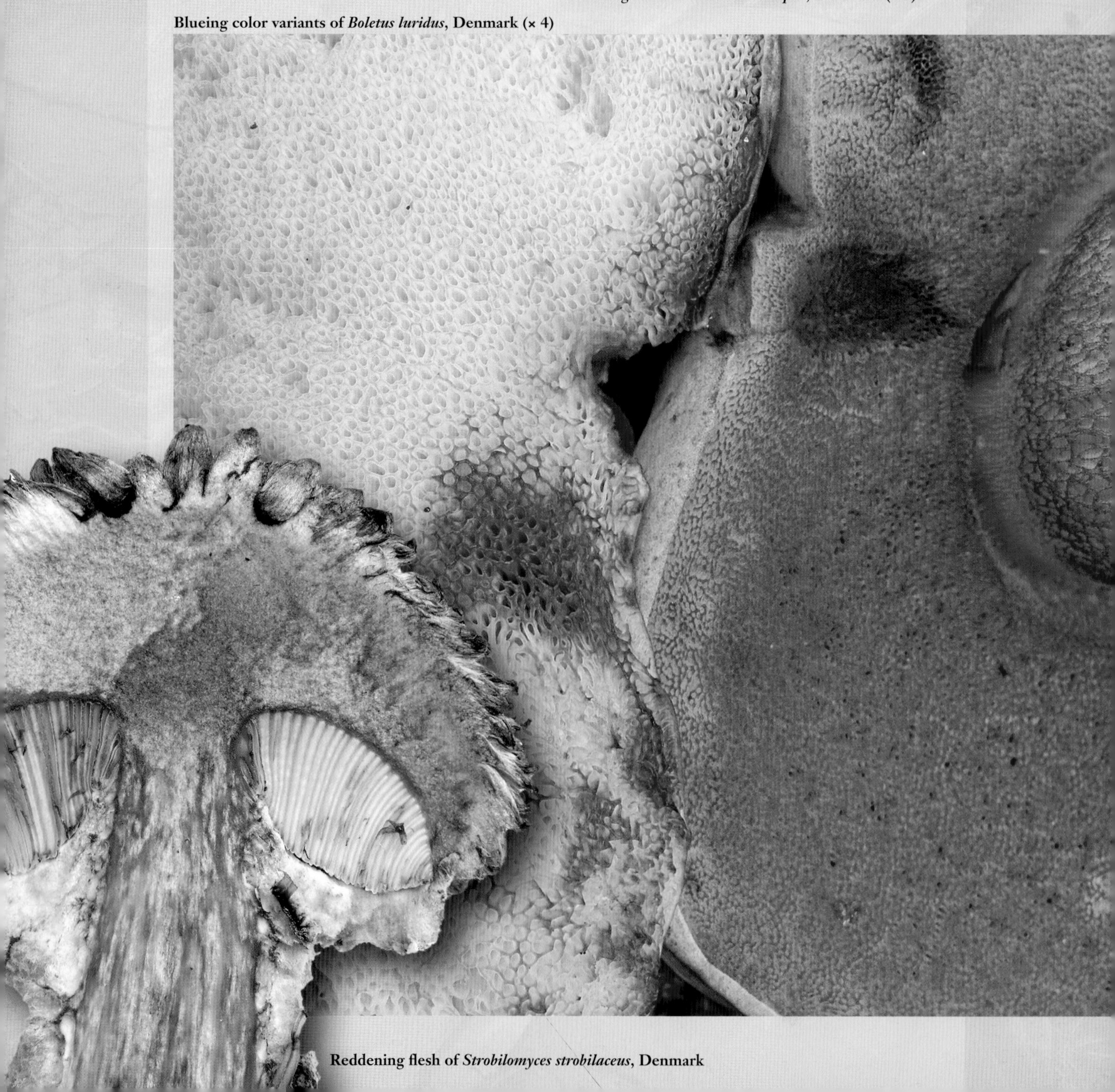

Reddening flesh of *Strobilomyces strobilaceus*, Denmark

Blueing flesh and tubes of *Gyroporus cyanescens*, Denmark (× 8)

Blackening *Leccinellum crocipodium*, Denmark (× 2)

Yellowing *Tylopilus virens*, Bhutan

Porcini, penny bun, cep, bolet cèpe, Steinpilz, steinsopp, Karl Johan: a favored bolete with many names. *Boletus edulis* is one of the most beloved of edible fungi: easy to recognize, common, and with an excellent taste.

The "species" as interpreted by most mushroom collectors is actually a species group. All members are edible and easily recognized by their white to yellow tube mouths that don't change color when touched, their pale, netlike pattern at the top of the stem, and their mild, nutty taste.

Boletus edulis, Denmark

Boletus edulis, Denmark (× 2)

Boletus edulis, Denmark (× 2)

The boletes are placed in many separate genera (and even in several families): *Aureoboletus*, *Austroboletus*, *Boletus*, *Boletellus*, *Gyroporus*, *Leccinum*, *Phlebopus*, *Strobilomyces*, *Suillus*, *Tylopilus*, and *Xerocomus*, to name just a few.

Characters such as the presence of veils, tube and spore color, and spore morphology are important in separating these genera.

Boletellus emodensis, Bhutan

Suillus luteus, a northern species introduced with *Pinus* seedlings to Ecuador (× 2)

Austroboletus fusisporus, Bhutan (× 4)

Aureoboletus gentilis, Denmark

Agarics

The agarics are recognized by the presence of *gills* beneath the cap. The cap can sit on top of a stem or it can grow directly from branches or herbs.

The basidia are placed on the sides of the gills and shoot their spores into the spaces between the gills. Thus the gills must have more or less vertical sides to give free passage to the falling spores.

The agarics are one of the largest groups of fungi, with more than five hundred genera and thousands of species worldwide. This is also one of the most eye-catching fungal groups, with fruiting bodies both tiny and huge.

Tiny *Mycena capillaris* on leaves, Denmark (× 8)

Hygrocybe psittacina, Denmark (× 16)

The fly agaric (*Amanita muscaria*), a poisonous, though not deadly, species, Denmark (× 3)

The features of the gills are among the most important characters of the agarics.

Gills may be *free* (not touching the stem) or *attached*. Attached gills may be attached very narrowly (*adnexed*) or broadly (*adnate*), or they may be more or less *decurrent*. An intermediate type is the *emarginate* gills, which first bend toward the cap but then turn and go slightly down the stem.

Pluteus romellii with free gills, Denmark (× 10)

Parasola plicatilis with free gills, Denmark (× 10)

Marasmius berteroi with almost free gills, Ecuador (× 10)

Entoloma serrulatum with adnexed gills, Denmark (× 10)

Mycena amicta with adnate gills, Denmark (× 12)

Psathyrella pseudogracilis with somewhat emarginate gills, Denmark (× 10)

The gills may be all possible colors, either from pigment in the tissue or from colored spores.

Cortinarius semisanguineus with emarginate gills, Denmark (× 10)

Cortinarius violaceus with emarginate gills, Denmark (× 5)

Leucopaxillus rhodoleucus with shortly decurrent gills, Denmark (× 10)

Hygrocybe miniata with shortly decurrent gills, Denmark (× 10)

Lentinus sp. with decurrent gills, Ecuador (× 3)

Arrhenia philonotis with decurrent gills, Denmark (× 8)

Close-up of *Mycena pterigena* with no veil, Denmark (× 200)

Some agarics develop their fruiting bodies and gills without any protection. Others have evolved various veils to protect the young structures from drought and predation by snails and insects.

A *universal veil* covers the whole fruiting body with slime, threads, or a membrane. When the fruiting body grows, the veil breaks. It is often possible to find veil remnants on the cap or as a sheath at the base of the stem.

The deadly poisonous deathcap (*Amanita phalloides*) emerging from its membranaceous universal veil, Denmark (× 3)

The mature deathcap the the universal veil appearing as a sheath (volva) at the base of the stem (× 1.5)

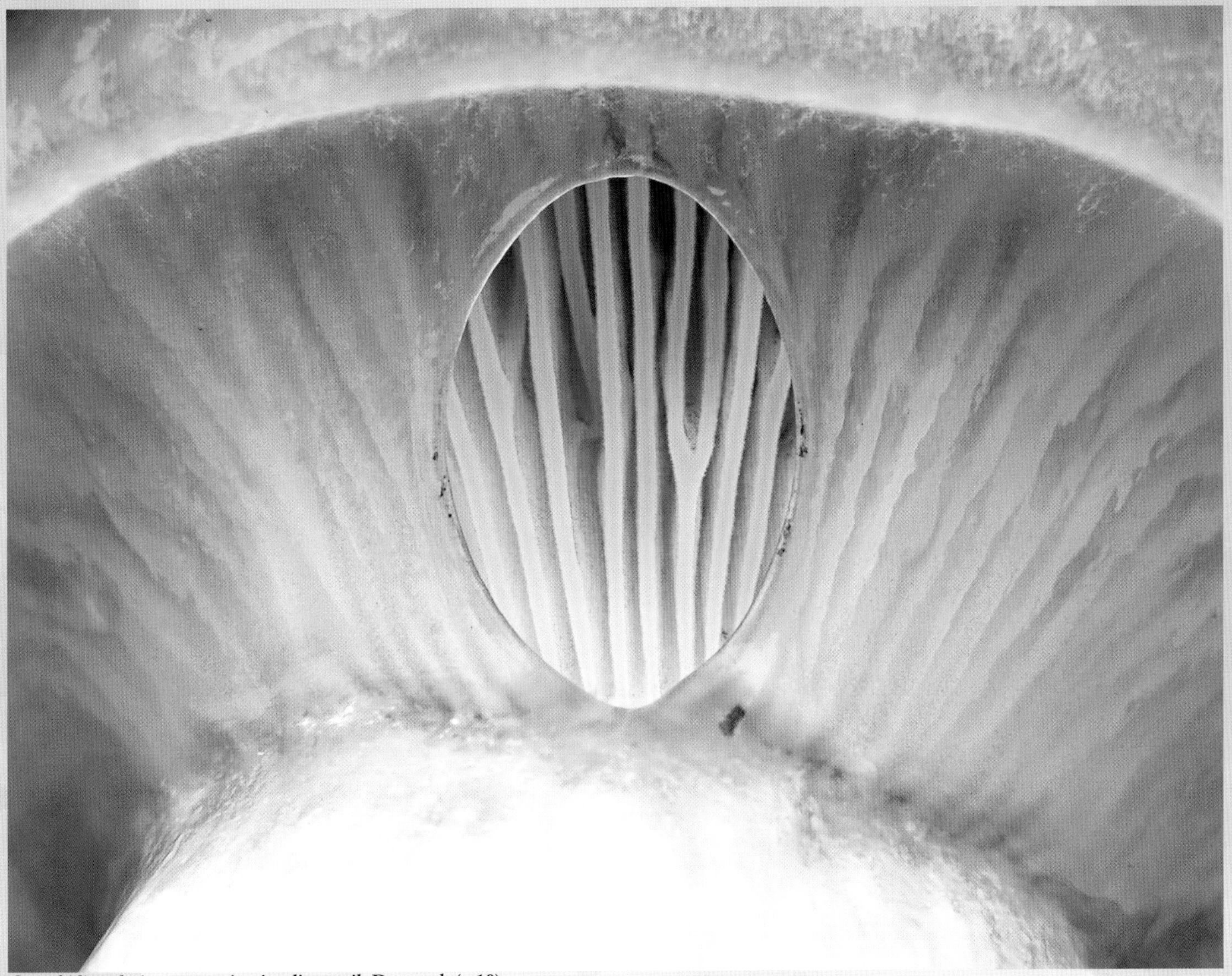

Gomphidius glutinosus opening its slimy veil, Denmark (× 10)

Cortinarius purpurascens with a cobweb-like veil, Denmark (× 4)

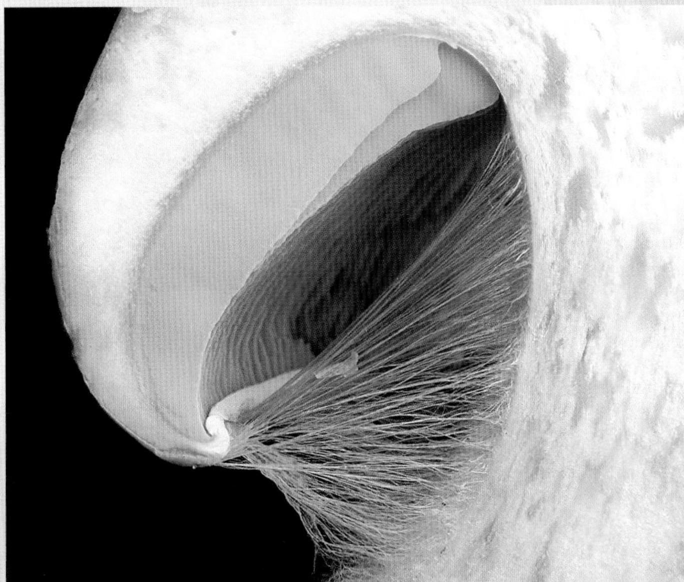

Cortinarius bergeronii with veil, Denmark (× 2)

A *partial veil* connects the stem to the margin of the cap, thus protecting the young gills. As the cap expands, the veil breaks. It may remain as a *ring* or as velar threads on the stem. Rings may also be formed by the universal veil og a mixture of both.

***Pholiotina vexans*, Denmark (× 10)**

***Amanita hemibapha*, Bhutan**

Mucidula mucida, Denmark

Agaricus arvensis, Denmark

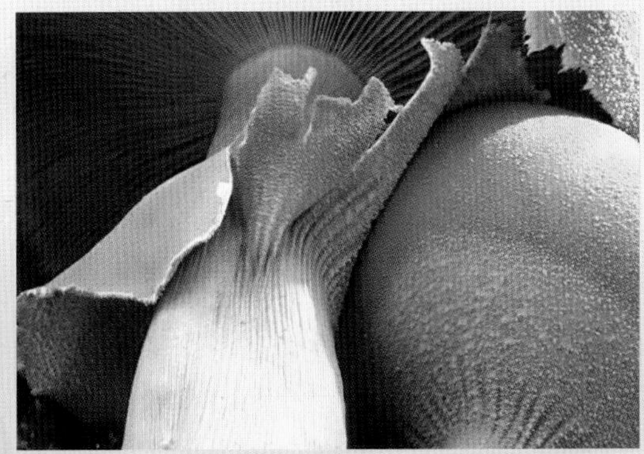

Phaeolepiota aurea, Denmark

The surface of the fruiting bodies—and especially the gills—may be covered by projecting cells called *cystidia* or thick-walled, pointed hairs called *setae*.

In some species, the edges of the gills are colored by colored cystidia. In other cases the edges may be slimy with mucus.

Marasmius hudsonii with both long setae and small cystidia on cap and stem, Northern Ireland (× 15)

Gills of *Coprinellus micaceus* covered by cystidia, Denmark (× 50)

Hygrocybe vitellina (×

Hygrocybe psittacina (× 10)

Despite the strong colors all pictures on this page show white-spored agarics. To observe the spore color, a spore deposit is needed: place the cap on a piece of paper for several hours, and a pattern of spores—in this case, white—may appear on the paper from the spaces between the gills.

Amanita muscaria
and spore deposit

The genera *Russula* and *Lactarius* constitute a special group of gilled fungi that are not closely related to the main bulk of agarics. One defining character is their brittle flesh, which will break without any longitudinal structure and is often likened to a dry cheese. In addition, species of *Lactarius* exude a milky fluid when cut or broken.

Most of these species form large, colorful fruiting bodies and are among the most important ectomycorrhizal fungi in most parts of the world.

Russula paludosa with brittle flesh, Denmark

Russula paludosa, Denmark

Lactarius uvidus, Sweden (x3)

Lactarius acris, Denmark (× 3)

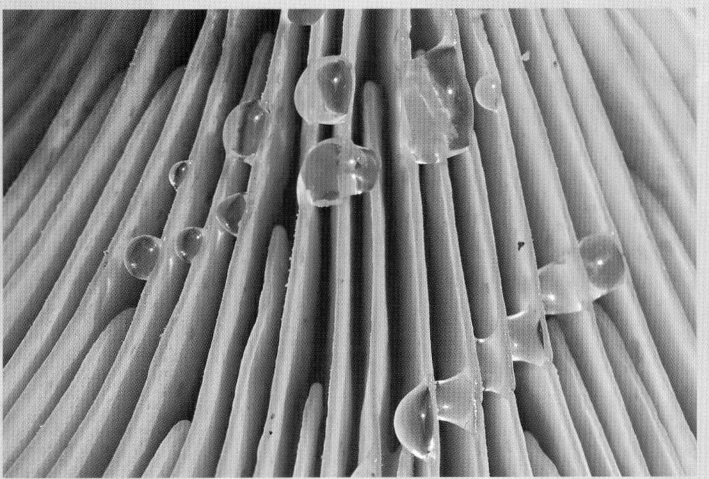
Lactarius helvus, Denmark (× 4)

Lactarius chrysorrheus, Denmark (× 2)

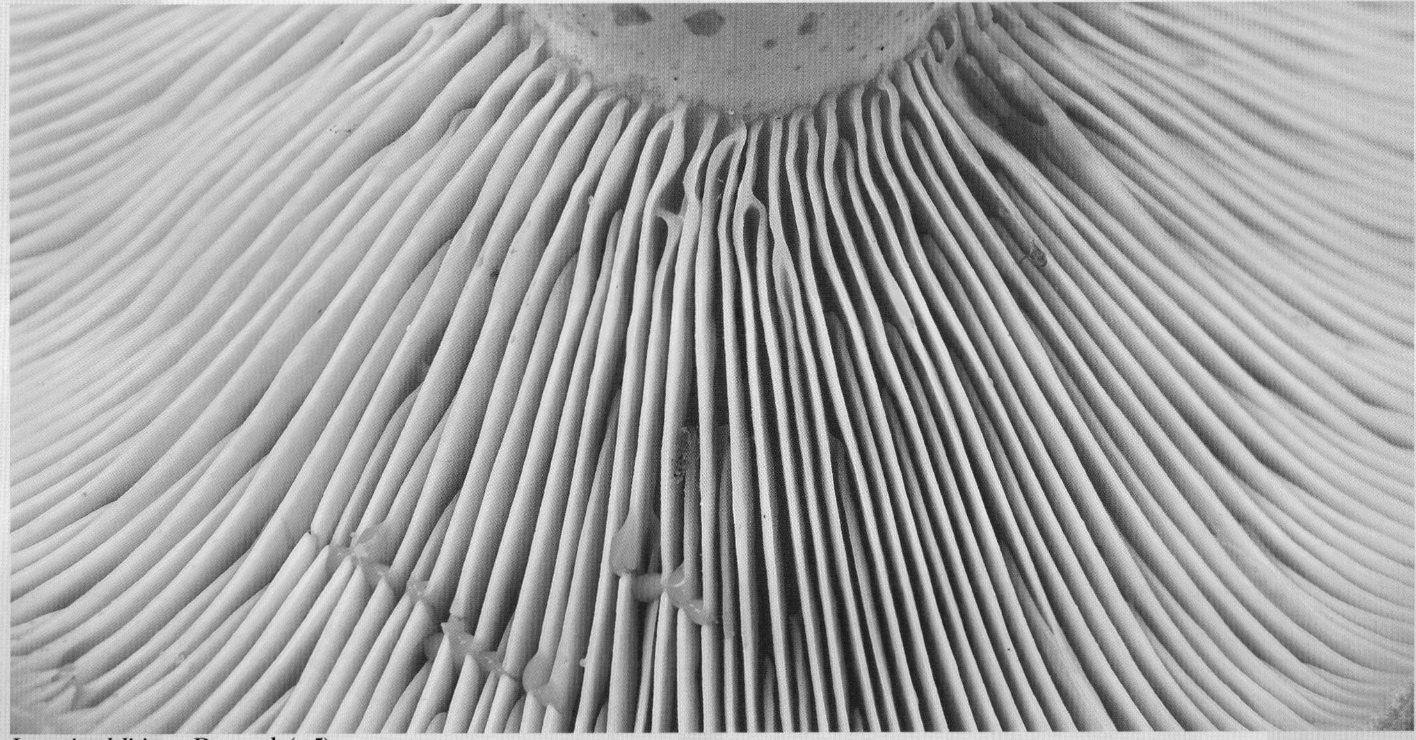
Lactarius deliciosus, Denmark (× 5)

Pleurotus ostreatus, Denmark (× 1)

Undescribed *Panus*-like fungus, Ecuador (× 2, TL)

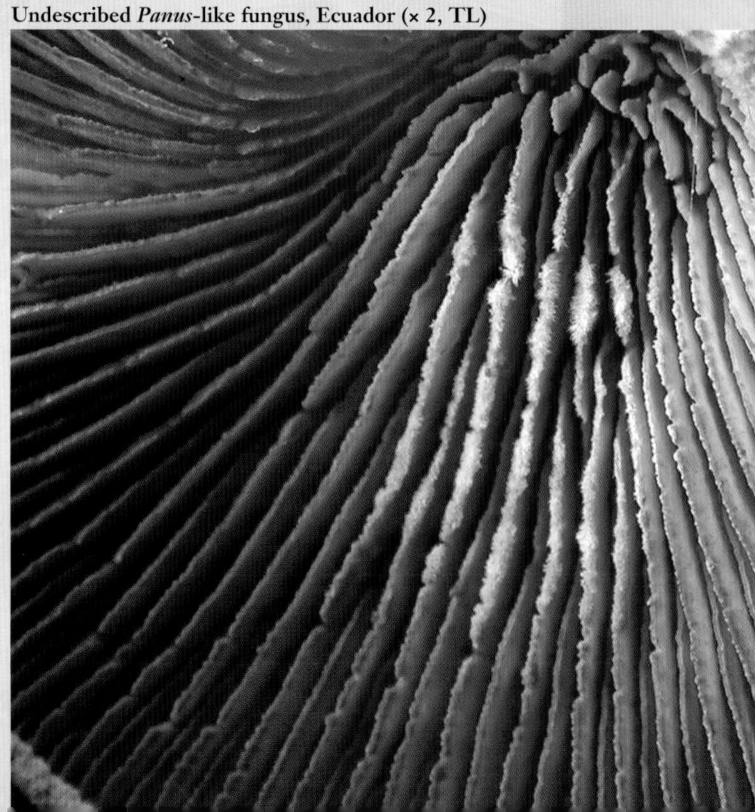

Some agarics have developed fruiting bodies without a stem. These are attached at the edge or top of the cap to wood or dead herbs. Typically these fruiting bodies are tongue- or kidney-shaped, with the gills continuing all the way to the substrate.

Crepidotus cinnabarinus, Denmark (× 15)

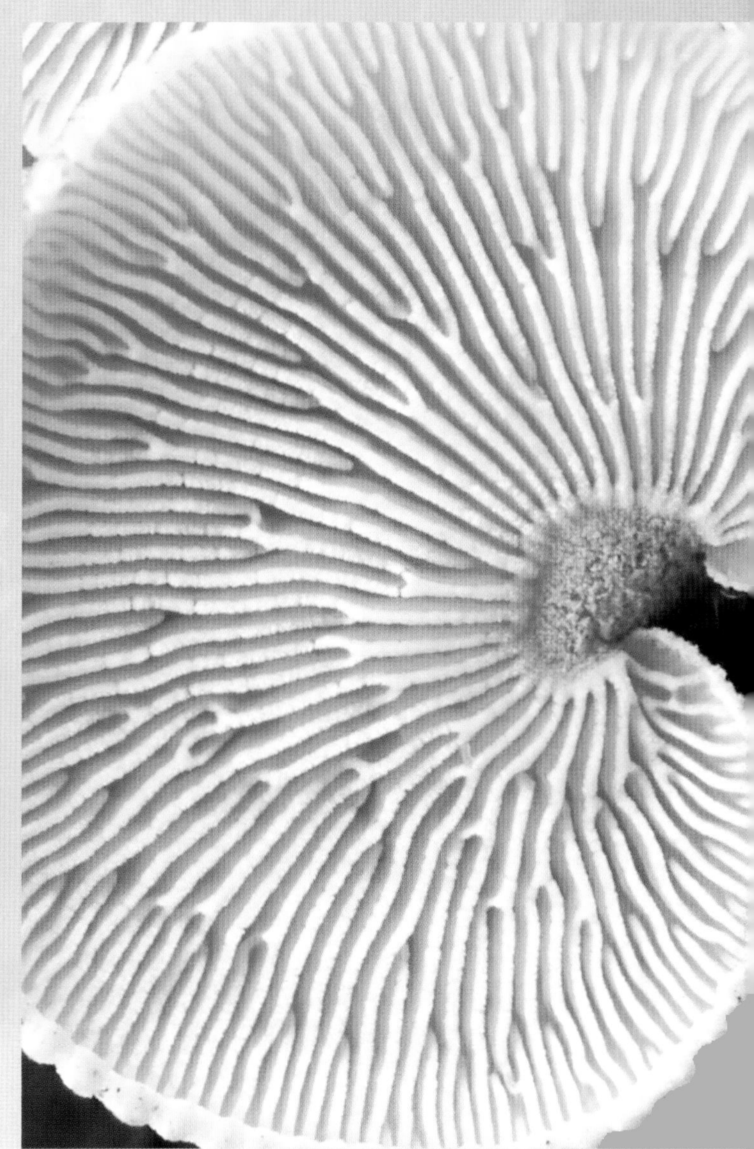

Panellus sp., Bhutan (×15)

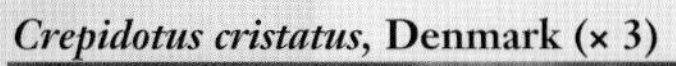

Crepidotus cristatus, Denmark (× 3)

Cyphelloid fungi

The cyphelloid fungi are small, cup-, bell-, tongue-, or tube-shaped Basidiomycota that hang from wood or dead herbs. Their fruiting bodies resemble those of the cup fungi (Ascomycota, see page 40-41), but in contrast to these they must point their opening downward to allow the weakly released basidiospores to fall free.

Flagelloscypha oblongispora, Denmark (× 25)

Lachnella villosa, Denmark (× 25)

Unnamed cyphelloid fungus, Ecuador (× 25)

Calyptella campanula, Denmark (× 15)

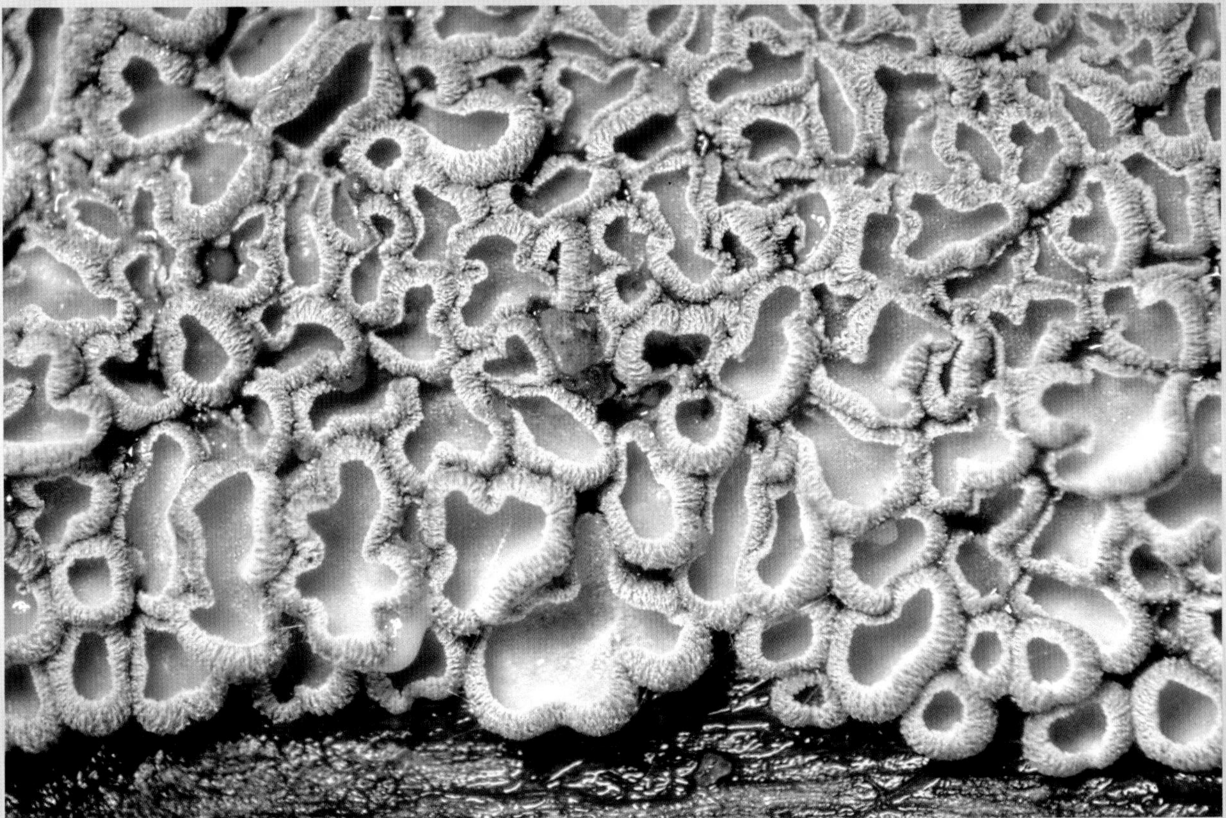

Merismodes anomalus (*Cyphellopsis anomala*), Denmark (× 20)

Henningsomyces candidus, Denmark (× 20)

When cyphelloid fungi fruit closely together, they may form meta–fruiting bodies by fusing numerous single fruiting bodies. This is the case with the beefsteak fungus (bottom), where each tube may be perceived as the remnant of a single but now fused cyphelloid fruiting body.

The same may be seen in the splitgill species, and even in more typical agarics such as *Marasmius bulliardii* (facing page), in which each gill lobe can be seen as a flattened, cyphelloid fruiting body.

The secret in constructing the complex agaric fruiting body may, indeed, lie in fusing smaller ones into a meta-structure.

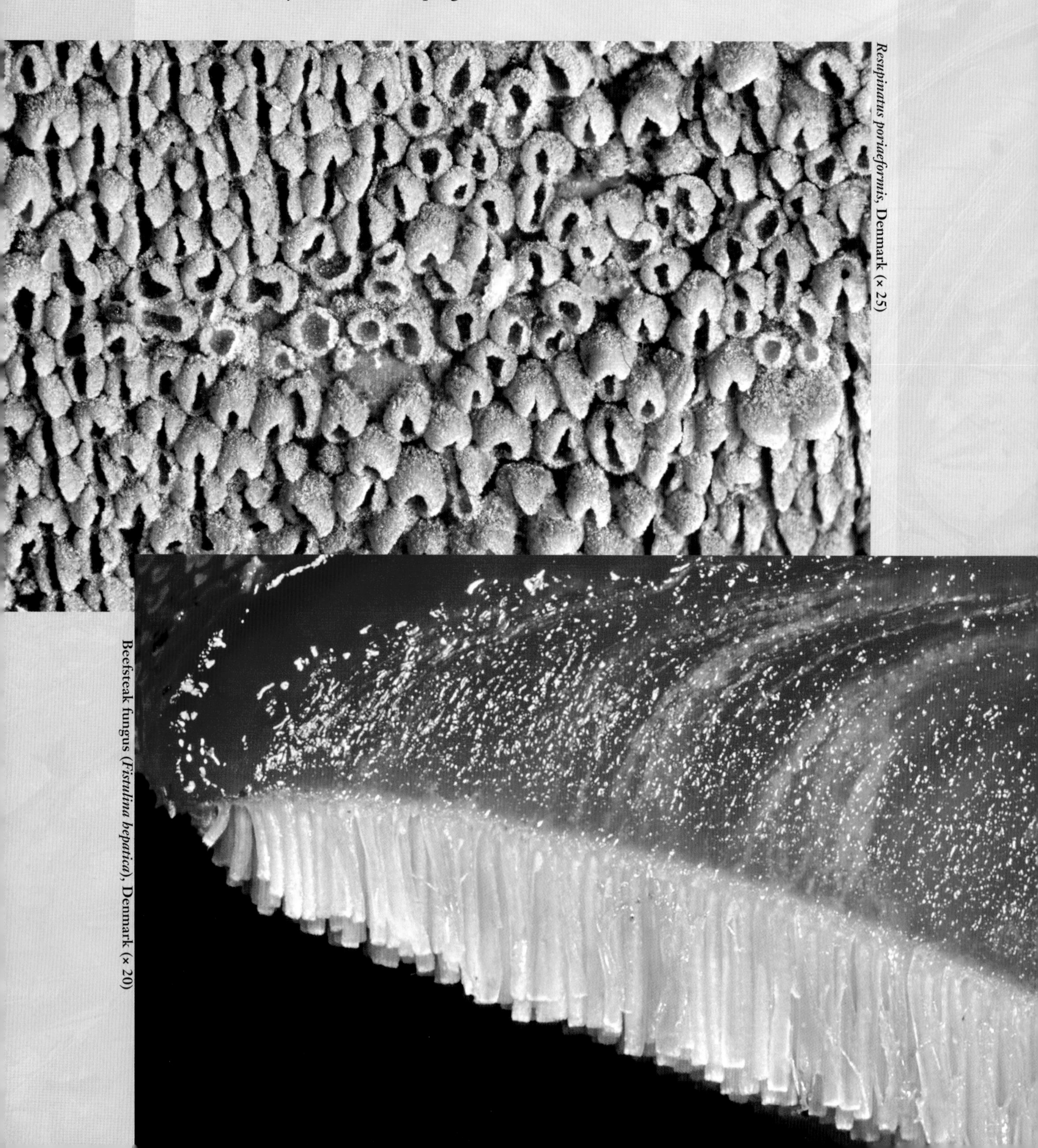

Resupinatus poriaeformis, Denmark (× 25)

Beefsteak fungus (*Fistulina hepatica*), Denmark (× 20)

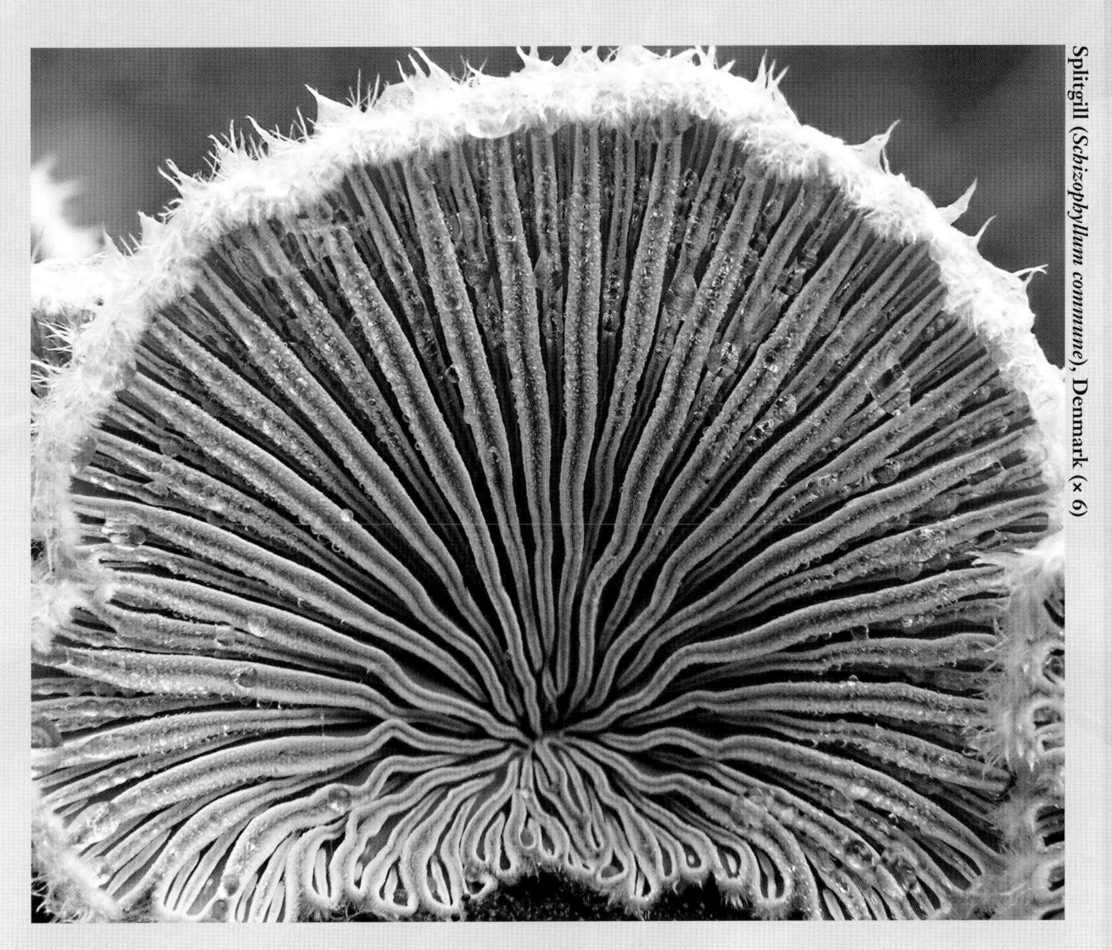

Splitgill (*Schizophyllum commune*), Denmark (× 6)

Marasmius bulliardii, Denmark (× 6)

Chanterelles

Chanterelles are characterized by a somewhat funnel-shaped fruiting body with a smooth or veined underside. They typically form ectomycorrhiza and are thus bound to forest ecosystems.

Because of their very characteristic fruiting bodies and unique tastes, the chanterelles are among the most highly praised of edible fungi. Growing ectomycorrhizal fungi commercially would, however, necessitate the cultivation of both trees and their fungal partners. As such a system has shown very complicated, all chanterelles that are sold are harvested in nature.

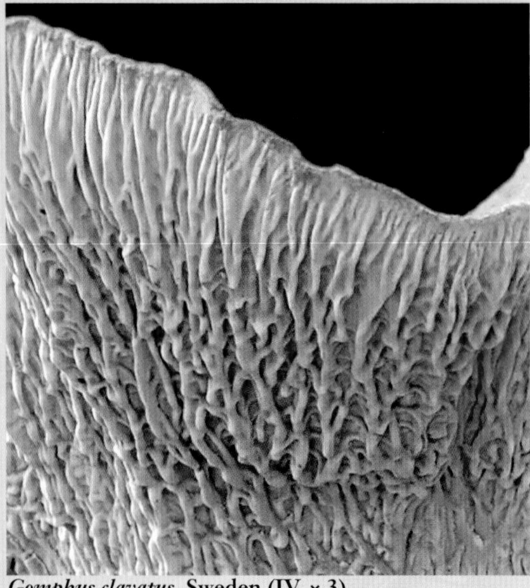

Gomphus clavatus, Sweden (JV, × 3)

Cantharellus pallens, Denmark (× 2)

Craterellus cornucopioides, Denmark (× 1.5)

Clavulina craterelloides, Ecuador (× 2)

Gomphus floccosus, Bhutan (× 4)

***Cantharellus cibarius* is the name most commonly used for the yellow chanterelles picked, sold, and eaten in temperate and subtropical parts of the world. As is often the case with these archetypal fungi, this name probably covers a range of closely related species with minor variations in morphology, ecology, smell, and taste.**

Cantharellus sp. on sale, Bhutan

The true Northern European *Cantharellus cibarius*, Denmark

Cantharellus sp. locally named *Cantharellus cibarius* but with aberrant tomentose cap and magnificent taste, Bhutan (× 6)

Tooth fungi

Tooth fungi have a lower cap surface covered by small teeth or spines, which are again covered by the hymenium. To enable the spores to disperse freely, the teeth are positioned vertically.

There are also fungi with teeth and spines among the corticioid fungi. These, however, form flat fruiting bodies, typically attached to dead wood (see page 106).

Hydnellum aurantiacum, Bhutan (× 10)

Hydnellum ferrugineum, Finland (× 3)

Hydnum repandum, Denmark (× 3)

Many tooth fungi, including species of *Hydnum*, *Bankera*, *Hydnellum*, *Phellodon*, and *Sarcodon*, form ectomycorrhiza with trees. *Auriscalpium vulgare* is an exception to the rule: it grows on the previous year's cones—and only those from pine trees.

Auriscalpium vulgare, Denmark (× 15)

Auriscalpium vulgare, Denmark (× 10)

Phellodon melaleucus, Denmark (× 2)

A number of the tooth fungi are important indicators of high biodiversity. Whereas the above *Phellodon melaleucus* is a generalist, the *Hydnellum* species below are extremely picky about where they grow: unpolluted forests with long-term continuity—and consequently high biodiversity—are preferred. Thus the presence of several *Hydnellum* species points toward an especially rich habitat.

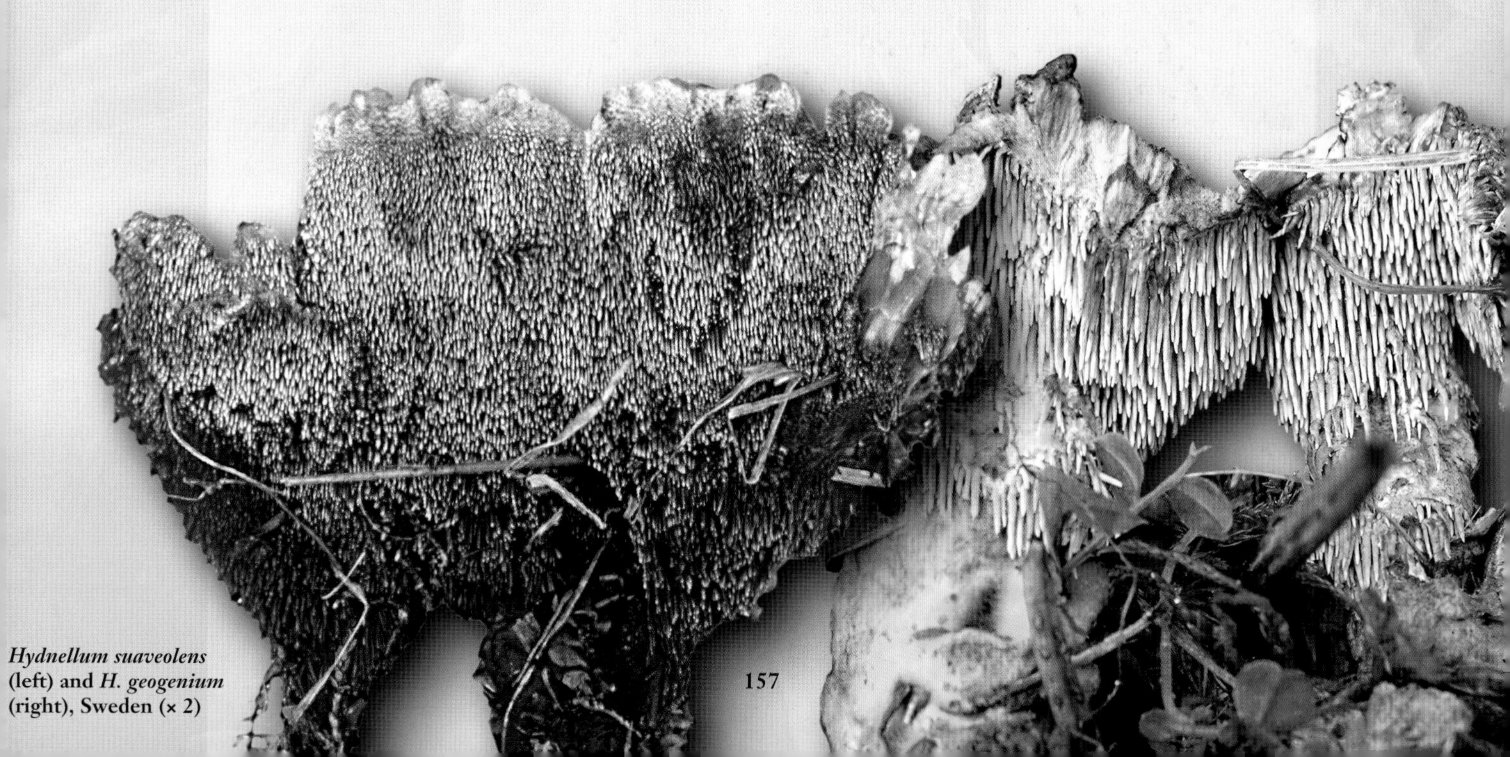

Hydnellum suaveolens (left) and *H. geogenium* (right), Sweden (× 2)

Hericium coralloides, Denmark

Also the wood-dwelling tooth fungi normally have a stem, but these grow more or less laterally attached, like the polypores. They are sometimes called the hericiaceous fungi. As many of these species demand long-term continuity and very large fallen or standing trunks, they are indicators of forests with especially high biodiversity.

Hericium erinaceus, Denmark

Clubs and corals

The clavarioid fungi are basidiomycota with club- or coral-shaped fruiting bodies. These often have a sterile base and produce spores on the remaining outer surface.

Club- and coral-shaped fruiting bodies are highly successful and have evolved numerous times in many different fungal lineages (see page 42-43).

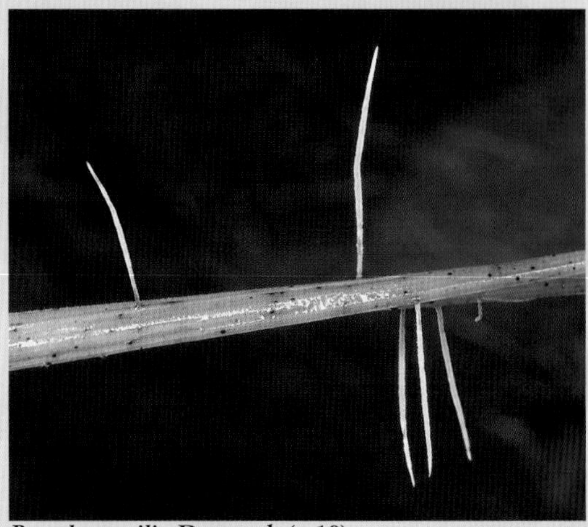

Pterula gracilis, Denmark (× 10)

Ramaria sp., Bhutan (× 4)

Clavariadelphus pistillaris, Denmark

Pterula gracilis—showing basidia with spores in the middle and cystidia at the base, Denmark (× 100)

Because the spore-producing tissue of clavarioid fungi is exposed to drought, rain, insects, and snails, the chance of producing viable spores may seem rather low. Also, larger, branched fruiting bodies—like those of *Ramaria*—have obvious problems dispersing spores from their central portions.

On the other hand, the construction of these fruiting bodies is remarkably simple, they may start producing spores very young, and apical growth may continue for many weeks, thus boosting spore production.

Typhula culmigena with 4-spored basidia on the club, Denmark (× 70)

Young fruiting bodies of *Ramaria pallida* just starting to produce spores, Denmark

Old *Ramaria pallida* with the inner portion dusted by brown spores that never escaped, Denmark

There is great variation in color and morphology among the club and coral fungi. Their fruiting bodies can be not only simple or branched, but also worm shaped, cylindrical, or trumpet shaped, or they can arise from *sclerotia* and be all possible colors.

Typhula gyrans, Denmark (× 4)

Clavaria incarnata, Denmark (× 2)

Clavulinopsis luteoalba, Denmark (× 2)

Typhula crassipes, Norway (× 5)

Clavicorona taxophila, Denmark (× 5)

Clavariadelphus truncatus, Sweden (× 0.5)

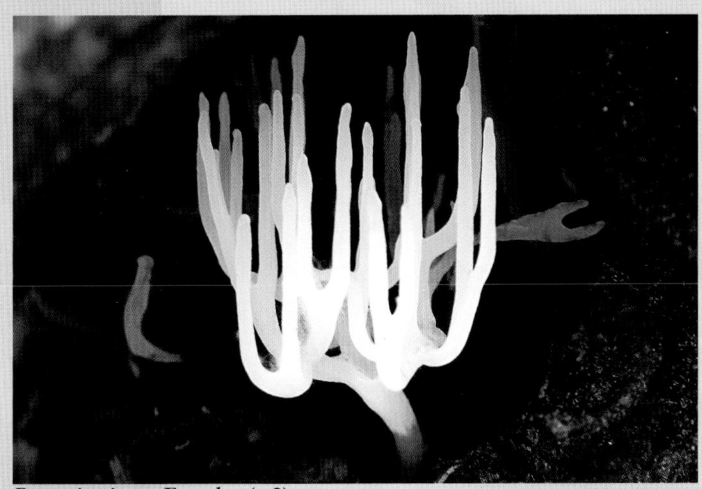
Ramariopsis sp., Ecuador (× 2)

Ramariopsis pulchella, Denmark (× 4)

Phaeoclavulina abietina, Denmark (× 2)

Ramaria sanguinea, Denmark (× 0.5)

Ramaria botrytis, Denmark (× 0.5)

Ramaria flavicingula, Sweden (× 0.2)

Pterula sp., Ecuador (× 4)

Different groups of clavarioid fungi have evolved in the different climatic zones of the world. Genera that are diverse in the temperate zones (e.g. the ectomyorrhizal *Ramaria*) may be scarcer in the tropics, where others dominate. In the Amazonian rainforest, the genus *Pterula* explodes with species; and genera like *Aphelaria* and *Scytinopogon*, which are not seen farther north, thrive.

Pterula sp., Ecuador (× 4)

Aphelaria tropica, Ecuador (× 3)

Scytinopogon sp., Ecuador (× 2)

Deflexula sprucei, Ecuador (× 8)

Physalacria sp., Ecuador (TL, × 8)

Mucronella calva, Denmark (× 16)

Mucronella sp., Bhutan (× 10)

In some cases clavarioid fungi point downward instead of reaching for the sky. An intermediate form is the weird *Deflexula sprucei*, with a main branch that points up and secondary branches that hang down. Species of *Physalacria* are somewhat similar, with erect stems and hanging heads, whereas *Mucronella* species go to the extreme with nicely hanging clubs.

The Cauliflower fungi (*Sparassis*) are another fungal type that does not fit properly into the general categories. These fungi have large, complex fruiting bodies with curly, flattened branches bearing the hymenium. They decompose dead wood and make a superb meal when cooked.

Sparassis crispa, Denmark

Gasteroid fungi

Gastro- is Greek for "stomach," and the gasteroid fungi are Basidiomycota that produce their spores inside—in the "stomach." They never actively shoot their spores from the basidia but rely on some outer force—like an animal or the splashing of raindrops—to get the spores airborne.

A typical example is the puffballs, which at first are massive and filled with basidia and spores. At maturity, a pore may form at the top and the spores—now a mass of dust—will be blown out when something touches the puffball.

Young *Lycoperdon perlatum*, Denmark (× 2)

Mature *Lycoperdon pyriforme* with a pore at the top Denmark (× 2)

Child dusting the forest with the spores of *Lycoperdon pyriforme*, Denmark

Boy enjoying the giant fruiting bodies of *Langermannia gigantea*, Denmark

Some of the most peculiar and charming fruiting bodies are found within the gasteroid fungi. The earthstars, for example, typically develop just below the soil surface. At maturity the fruiting body splits open and the whole structure is lifted by the triangular flaps while exposing the inner "puffball" containing the powdery spores.

Geastrum quadrifidum, Denmark (× 2)

Geastrum triplex, Denmark (× 2)

Geastrum coliforme with many openings, Ecuador (× 1)

Geastrum striatum with furrowed opening, Denmark (× 2)

Tulostoma exasperatum, Sabah, Malaysian Borneo (TL, × 3)

Chlorogaster dipterocarpi, Sabah, Malaysian Borneo (TL, × 2)

Like the earthstars, species of *Tulostoma* develop underground and emerge with stalked fruiting bodies when mature. As often happens, this lifestyle has been invented several times, as shown in the newly described but unrelated genus *Chlorogaster* with a similar lifestyle.

Phallus impudicus with witch's eggs, Denmark

Section through a witch's egg – the juvenile state of a stinkhorn, in this case a *Phallus multicolor*, Ecuador

One of the most bizarre groups of fungi contains insect-dispersed, foul-smelling species related to the stinkhorns (*Phallus*): devil's fingers, cage fungi, stink cigars, and others. These attract insects with a strong smell similar to that of flowers, feces, or carcasses. The insects then eat the blackish spore mass and thus disperse the fungi.

Phallus sp., Ecuador (× 2)

Clathrus archeri, Denmark

A stink cigar—*Staheliomyces cinctus*, Ecuador (× 2)

Open *Crucibulum crucibuliforme* with lens-shaped peridioles, Denmark (× 20)

Young, still-covered *Crucibulum crucibuliforme*, Denmark (× 4)

Bird's nest fungi are yet another weird type of fungus. Here the basidia and spores are placed inside small packages called *peridioles.* When the fruiting body matures, the lid covering the small cup disintegrates, allowing raindrops to splash into the cup and eject the peridioles. In some species the peridioles have a sticky surface or a small thread to allow them to better adhere to nearby vegetation.

Cyathus olla with gray peridioles, Denmark (× 15)

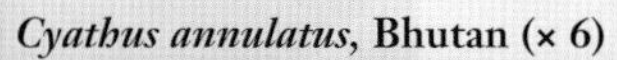

Cyathus annulatus, Bhutan (× 6)

Tremella mesenterica, Denmark (× 4)

Different types of basidia (≈ × 700)

Jelly fungi

The jelly fungi include Basidiomycota with gelatinized fruiting bodies. Under the microscope these are characterized by "strange" basidia, which either look like tuning forks or have transverse or longitudinal cell walls. The basidia are deeply immersed in the gelatinous fruiting body. In this way they become highly resistant to drought. At maturity, long sterigmata appear at the surface of the fruiting body to produce the spores.

Many jelly fungi are parasites on either plants or other fungi.

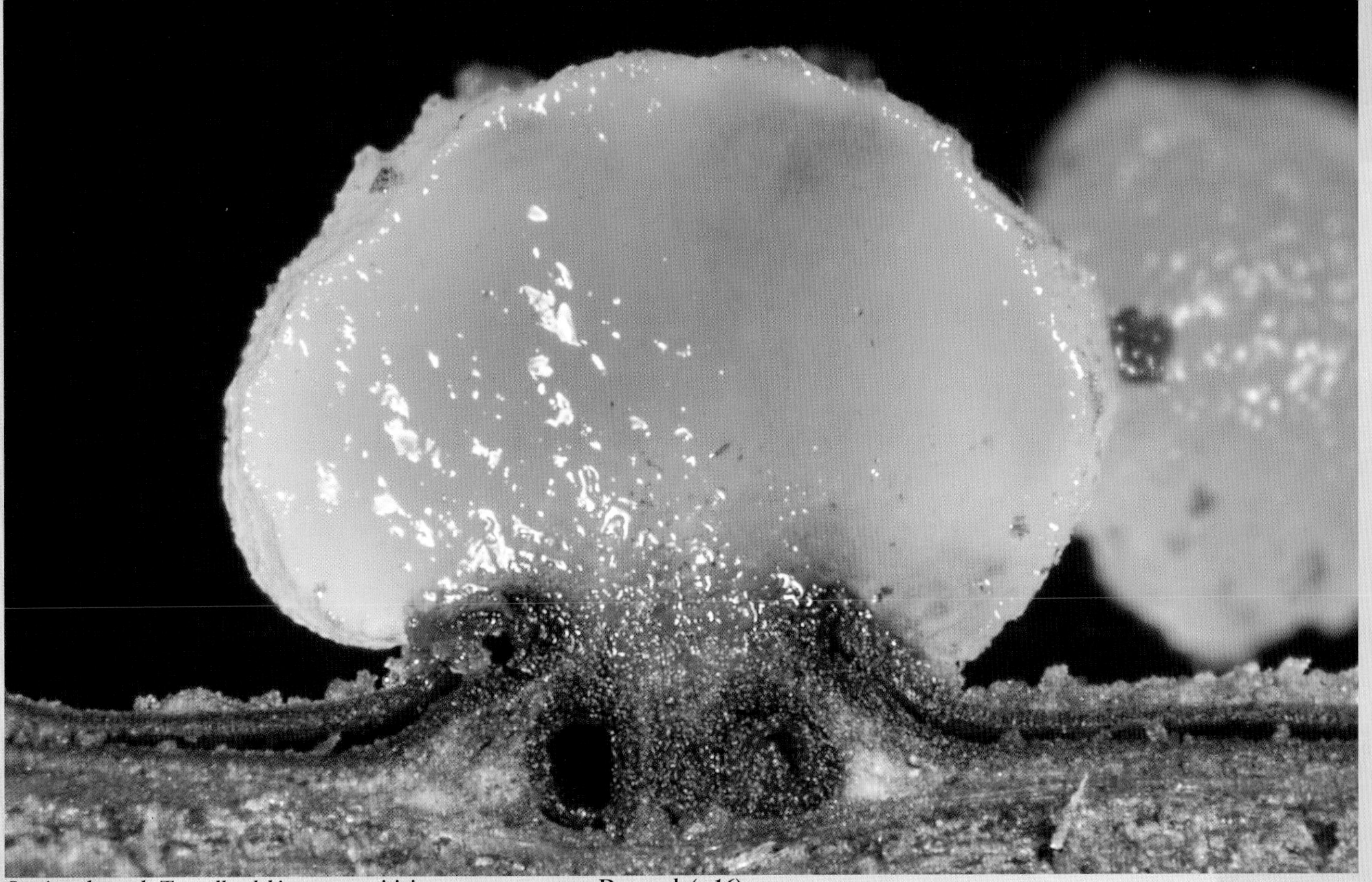

Section through *Tremella globispora* parasitizing a pyrenomycete, Denmark (× 16)

***Exidia plana*, Denmark**

Pseudohydnum gelatinosum, Denmark (× 2 and 10)

Elmerina holophaea, Bhutan (× 4)

Tremellodendropsis sp., Ecuador (× 1)

Calocera sp., Bhutan (× 4)

You can find most the major types of basidiomycote fruiting bodies among the jelly fungi. "Tooth fungi," "polypores," "coral fungi," and "club fungi" are all present—but with more or less gelatinized flesh and unusual basidia.

Auricularia delicata, Ecuador

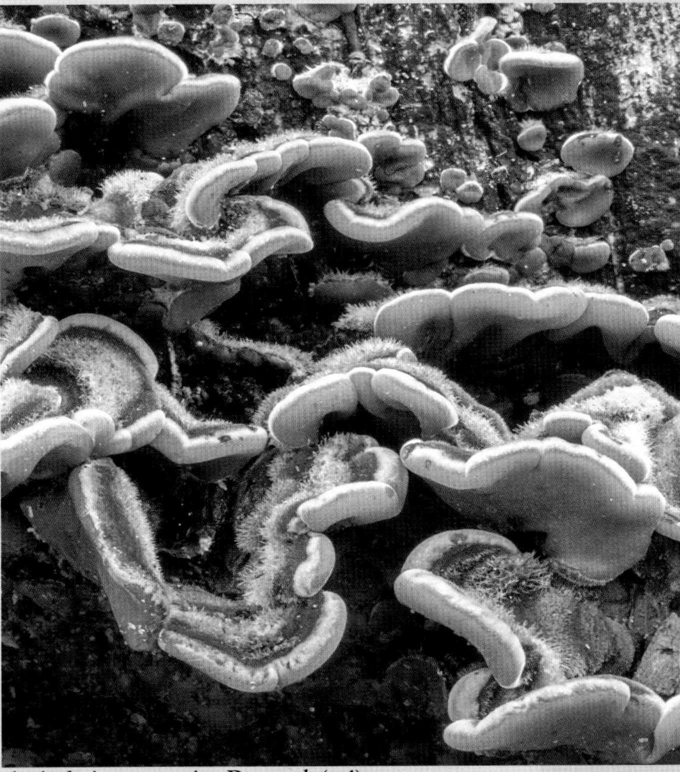

Auricularia mesenterica, Denmark (× 4)

The only jelly fungi of culinary importance are the species of *Auricularia* and the snow-white *Tremella fuciformis*. Auricularias are grown and sold dried under names like "black fungus," "cloud ear fungus," or just "wood ear." They are supposed to have medicinal properties and may enrich food with a certain cartilaginous texture but are virtually tasteless.

Auricularia auricula-judae, Denmark

Rusts and smuts

The rusts and smuts do not form real fruiting bodies. They are, however, remarkable and highly visible in nature.

These fungi are all parasites on living plants. A rust fungus typically infects its host in spring. Soon thereafter it forms *spermagonia* and *aecia* on the leaf surface. The spermagonia may exude nectar to attract insects, facilitating cross-fertilization by means of small spores called *spermatia*. Later the rust may jump to another host to form *uredinia* and eventually *telia*, from which sexual spores will grow the following spring.

Rust fungi can do great damage to agricultural crops.

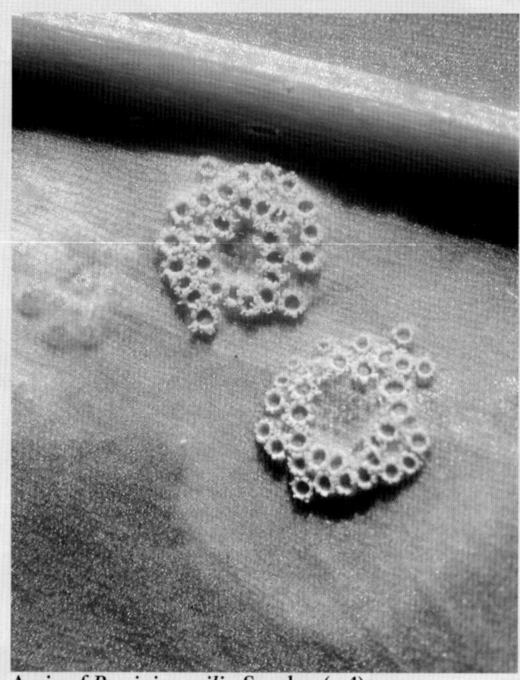

Aecia of *Puccinia sessilis*, Sweden (× 4)

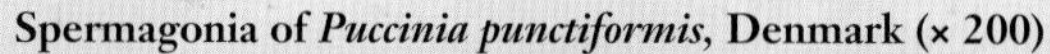

Spermagonia of *Puccinia punctiformis*, Denmark (× 200)

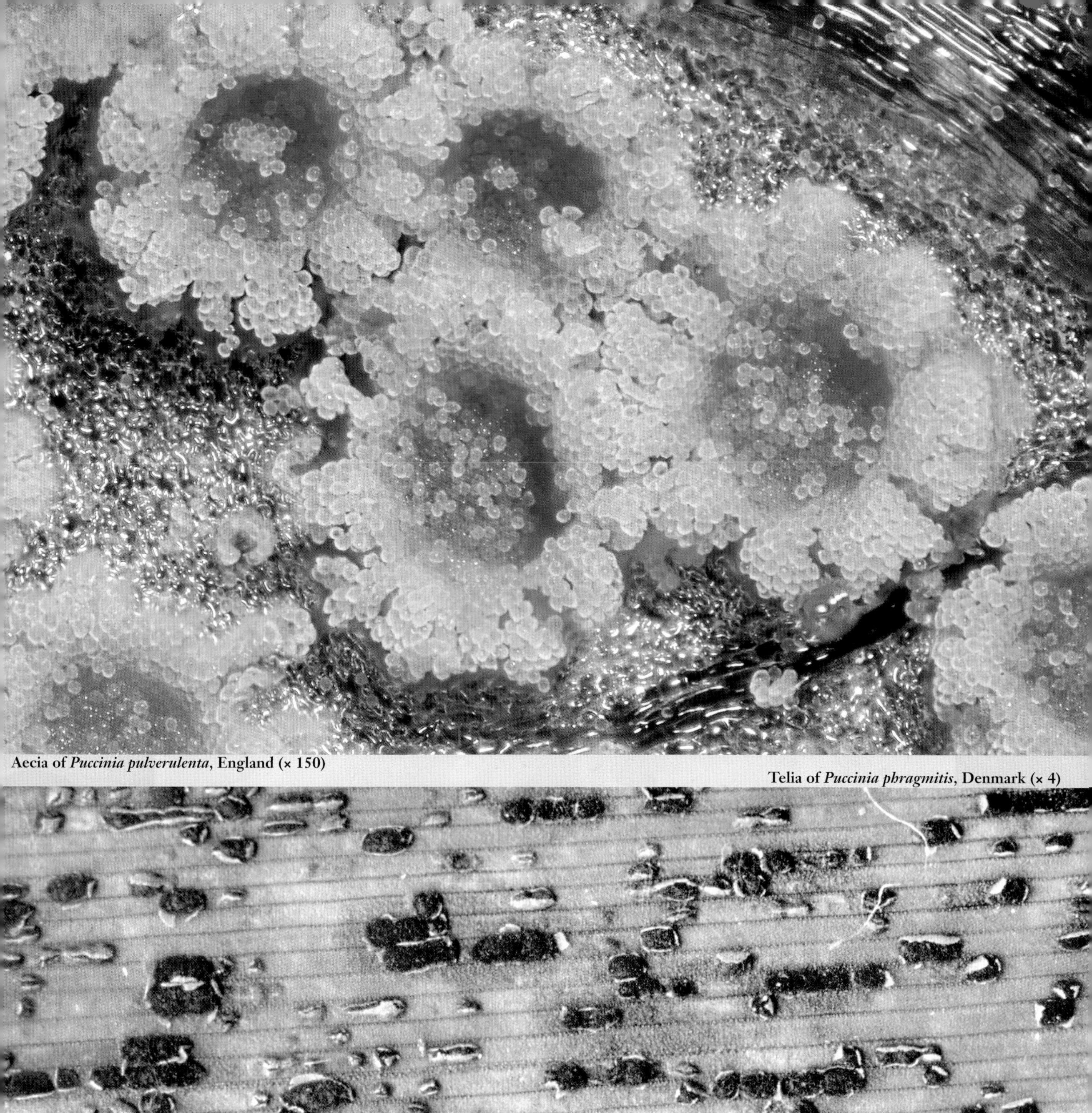

Aecia of *Puccinia pulverulenta*, England (× 150)

Telia of *Puccinia phragmitis*, Denmark (× 4)

Aecia of *Chrysomyxa ledi*, Finland (× 20)

Telia of *Gymnosporangium clavariiforme*, Denmark (× 3)

The smut fungi live less eye-catching lives than the rusts. Most of the year they dwell invisibly inside plants, but when ripe seeds would normally appear, a mass of blackish smut spores develops instead.

The smuts cause very serious damage to many crops, including cereals. A single species, *Ustilago maydis*, which parasitizes corn, forms large structures and has been eaten as a delicacy in Mexico since the time of the Aztec culture.

The smut *Ustilago maydis*, Mexico (DA)

The smut *Ustilago tritici* (also called *U. hordei*), Denmark (× 2)

The Zygomycota and other groups

Besides the Ascomycota and Basidiomycota, there are six more groups of fungi. Of these only the Zygomycota and the Glomeromycota form structures easily seen by the naked eye.

The Zygomycota are a form group, and with the exception of the genus *Endogone*, they never form real fruiting bodies. They do, however, form very conspicuous anamorphs—for example, *Entomophthora* and *Pandora*, which attacks living insects; *Kickxella* and *Pilobolus*, which form on dung; and *Mucor* and *Rhizopus*, which form on foodstuffs and other substrates (page 30).

Some species of the Glomeromycota may form truffle-like structures in soil. All members of the group plays an important ecological role forming A-mycorrhiza (page 216)

The four remaining phyla are all microfungi and are usually found and observed only by specialists.

Pandora dipterigena, England (× 10)

Entomophthora muscae, Denmark (× 16)

Ascomycota (65.000)

Chytridiomycota (700)

Basidiomycota (32.000)

Neocallimastigomycota (20)

Blastocladiomycota (179)

Dikarya

"Zygomycota" (1.100)

Microsporidia (1.500)

Glomeromycota (170)

Kickxella alabastrina, Denmark (× 60)

The asexual structure of *Pilobolus* bears a black spore-containing sporangium at the top. It can actively shoot this several meters away with velocities as high as twenty meters per second!

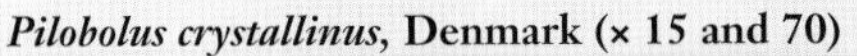

Pilobolus crystallinus, Denmark (× 15 and 70)

Fungal ecology

Fungi do not have chloroplasts. Thus they are unable to make their own energy but rely on energy (sugars) made by plants and algae. There are two main strategies for obtaining these sugars: *decomposing* organic material formed by plants or getting sugars directly from plants or algae by *parasitism* or *mutualistic symbiosis*—for example, by forming mycorrhiza or undergoing lichenization.

Decomposers

Most fungi are decomposers. Initially the nutrients in a growth substrate (for example, in dead wood, stems, or leaves) are bound in large, inaccessible molecules. To incorporate these nutrients into its growing mycelium, the fungus has to break the molecules into smaller units. This is done using *enzymes*.

Enzymes are proteins that facilitate specific chemical reactions. Some enzymes may start breaking down the cellulose in the plant cells while others will take care of the lignin. Depending on their enzyme systems, different fungi have the ability to degrade different substrates.

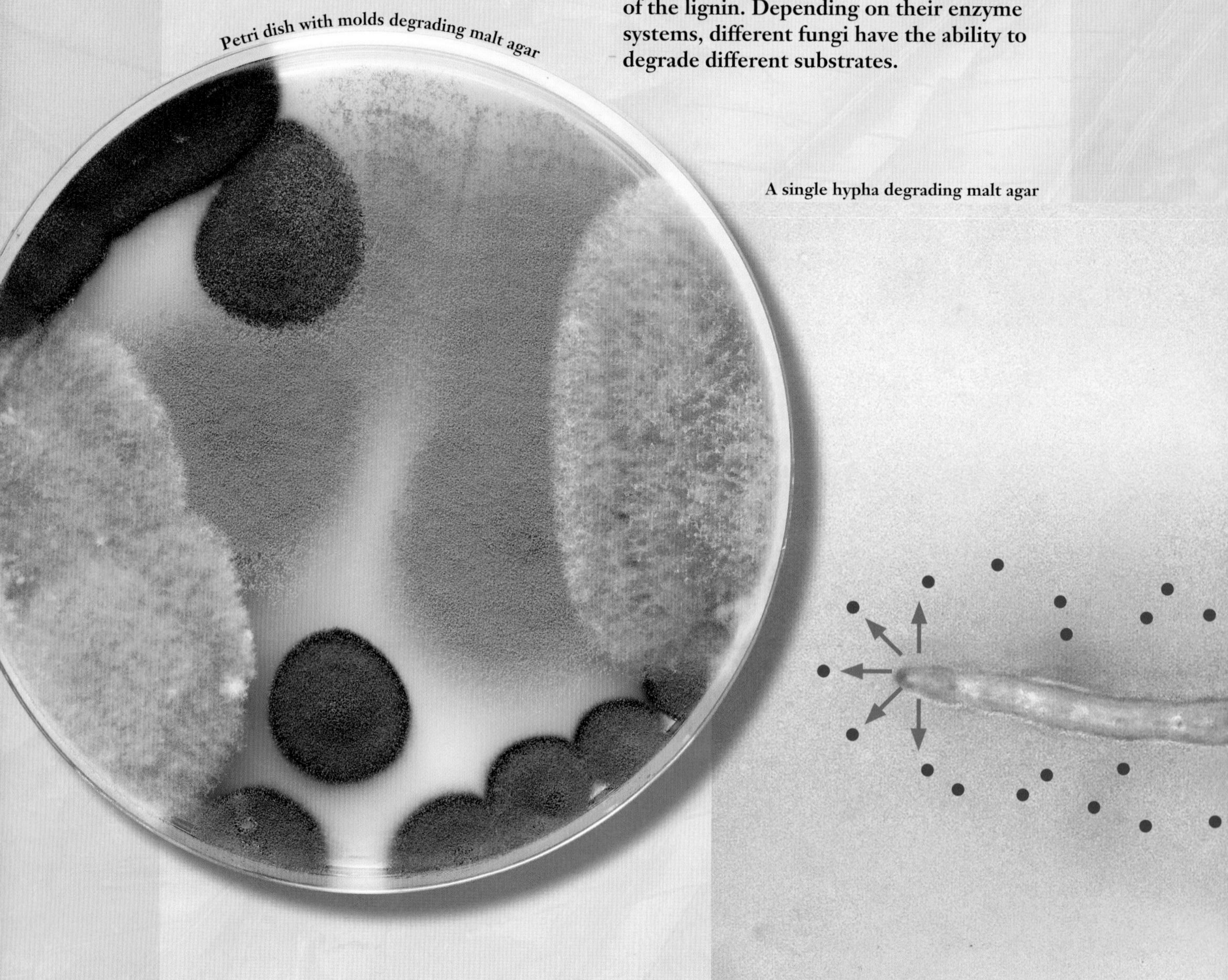

Petri dish with molds degrading malt agar

A single hypha degrading malt agar

When a hypha grows, vacuoles at its very tip will empty enzymes into the surroundings (below, orange arrows). These enzymes (dots) will then start degrading the growth substrate into smaller molecules. After degradation, the small molecules are able to diffuse back into the hypha, providing the fungus with nutrients and energy (green arrows).

Growing hyphal tip with vacuoles seen with an electron microscope (VW)

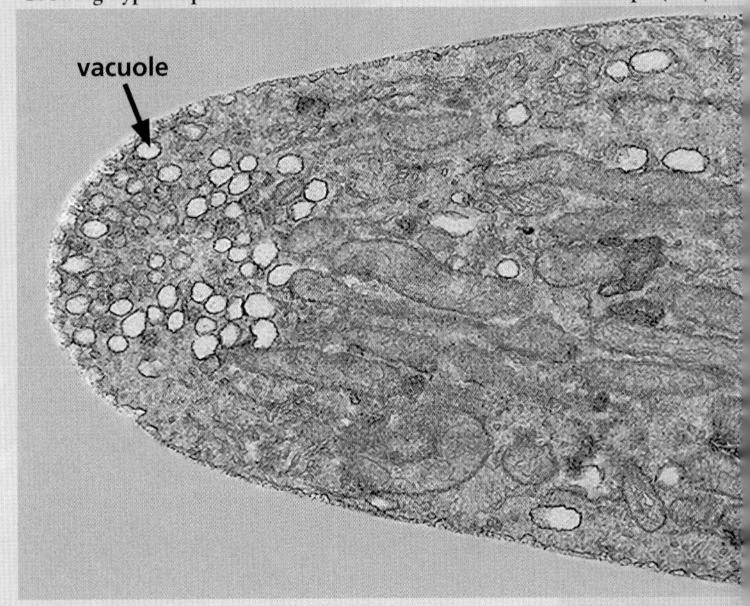

Orange arrows: release of enzymes; dots: enzymes; green arrows: nutirents diffusing back into the hypha

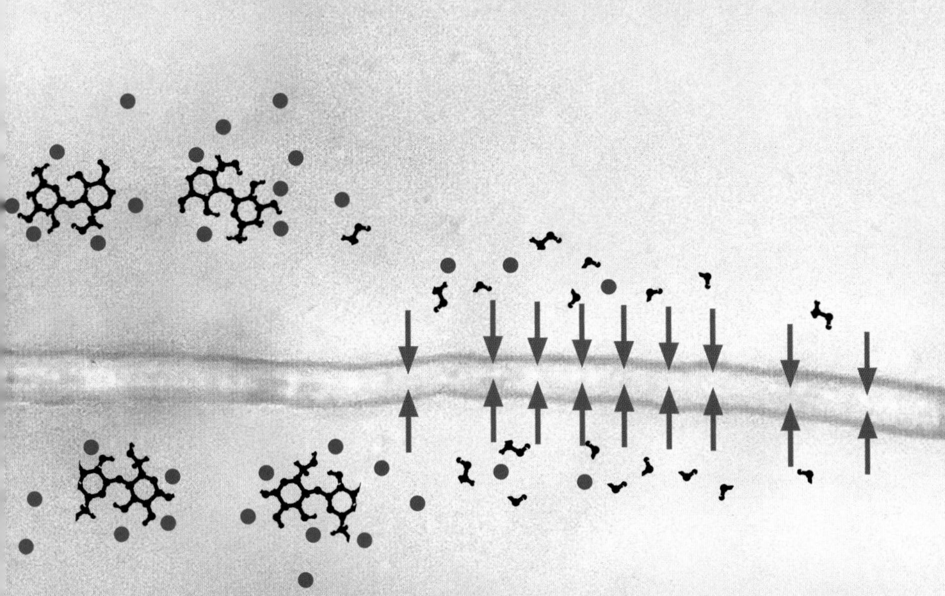

Mycena capillaris, Denmark (× 15)

Decomposers of debris

When a temperate deciduous forest drops its leaves in autumn, many tiny fruiting bodies of *Mycena*, *Marasmius*, and similar fungi appear. These are the pioneer decomposers, competing to be first to colonize the newly fallen leaves.

These fungi typically form rather small mycelia, inhabiting only a few leaves or coniferous needles, or even just a small part of one.

Some of the smallest fungal mycelia with functioning fruiting bodies are found on pine needles, where several *Lophodermium pinastri* individuals may compete over parts of the same needle, barricading each other with black shields or zones.

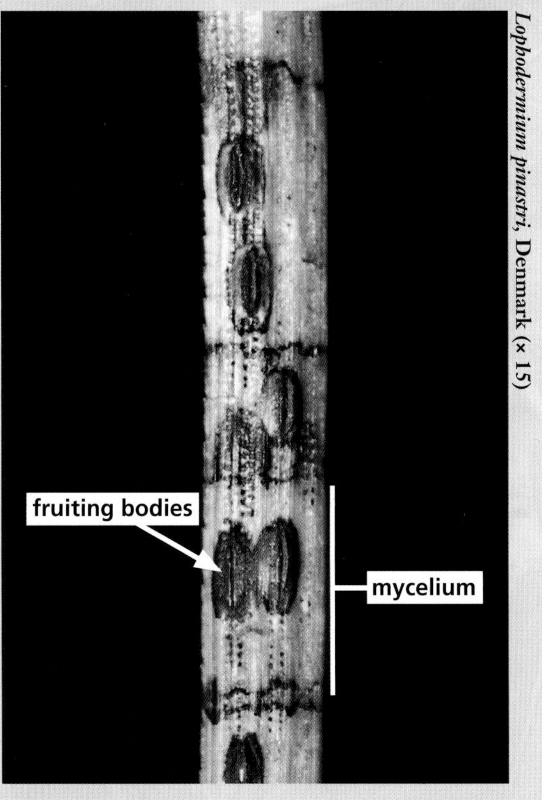

Lophodermium pinastri, Denmark (× 15)

Marasmius sp. weaving the rainforest debris together with its black horsehair-like cords, Ecuador (× 10)

In many forest ecosystems, not only are these fungi responsible for most of the litter decomposition, they also help stabilize the litter layer. Like a thread joining cloth into a dress, they sew the litter layer together with mycelia and hyphal cords. The result is a stable mat of debris that is likely to stay in place during the next rain. This is especially important in the tropical rainforest, where erosion is always a major threat.

Many decomposers of debris will not make do with a single needle or leaf but will instead form extensive mycelia that decompose a variety of dead organic remnants. Over several years, this type of fungus may use most of the available substrate in a central area and thus expand into a ring-shaped mycelium called a *fairy ring*.

Fairy rings are found in both forests and open land but are clearly most conspicuous in lawns and fields.

Gymnopus peronatus, Denmark

An 8-meter fairy ring of *Clitocybe nebularis* with approximately 1000 fruiting bodies weighing several hundred kilograms, Denmark

Lepista nuda, Denmark

Some grassland fungi form fairy rings that either kill the vegetation or make it grow more lushly. Thus the mycelia of these fungi can be seen even when no fruiting bodies are visible.

The most spectacular of these is probably *Leucopaxillus giganteus*. Not only does it form large fruiting bodies, it also creates dead zones in the vegetation that are thirty to forty centimeters wide, making the fairy rings easy to spot even in aerial photographs.

Fairy rings, including a 35 m ring of *Leucopaxillus giganteus* on an aerial photograph, Denmark

The above fairy ring confirmed on the ground—with fruiting bodies hidden in the tall grass (right)

Armillaria ostoyae, Denmark

An even more impressive fungal mycelium is that of a giant *Armillaria ostoyae* in the Blue Mountains of Oregon, dubbed "The Humongous Fungus." It covers an area of close to ten square kilometers, almost three times the size of Central Park in New York.

The mycelium—most likely a number of separate clones originating from the same source—is estimated to be between two thousand and eight thousand years old. This places this specimen of *Armillaria ostoyae* in the list of the top ten largest and oldest organisms in the world!

The Humongous Fungus is three times the size of Central Park, New York (© Martin St-Amant/Wikipedia)

Wood-decaying fungi

The main components of wood are cellulose and lignin. These structurally stable compounds are quite difficult to decompose, but fungi manage to do the job.

Some fungi are equipped with enzymes that can decompose cellulose. These cause a brown rot characterized by brown wood that may fall apart into cubical pieces (facing page).

Other fungi also have an enzyme system that can handle the more difficult lignin. These cause a white rot, named after the loose-textured, whitish wood left after the breakdown process (below).

Strongly decayed, white rotted beech trunk, Denmark

White rot in beech, formed by *Ceriporiopsis gilvescens*, Denmark

Brown rot in a coniferous trunk, formed by *Fomitopsis pinicola*, Bhutan

Shield zones and other structures in wood during and after fungal decomposition, Denmark

In the process of degrading wood, fungi may change it in other ways than just making it white or brown. More striking colors may appear, depending on the fungi involved. Also, dark or colored shield zones are often formed where the individual mycelia meet.

The shields seal the mycelia from each other, preventing a hostile takeover. Logically, the single mycelia mostly follow the structure of the wood, forming long, cigarlike shapes. This can be demonstrated by slicing a log and tracing the mycelia in the wood.

Students tracing, color coding, and reconstructing the more or less cigar-shaped mycelia inside a rotten log

Fungi on dung

Dung must also be decomposed, and fungi are, together with small animals, the main actors. Within the first two weeks after dung is dropped, species of the Zygomycote genus *Pilobolus* are already prepared to form asexual spores in spectacular sporangia. Other species quickly follow, first other asexual fungi and cup fungi, later flask fungi, agarics, and others.

Both generalists, which degrade many kinds of dung, and specialists, which are specific to the dung of geese, deer, cattle, and so forth, will be present. In total, species inhabiting dung number in the thousands.

Ascobolus sacchariferus, Denmark (× 35)

Lasiobolus intermedius, Denmark (× 40)

Cheilymenia granulata, Denmark (× 4)

Penicillium claviforme, Denmark (× 25)

Pilobolus crystallinus, Denmark (× 40)

Peziza sp., Sweden (× 4)

Sordaria fimicola, Denmark (× 80)

Coprinopsis stercorea, Denmark (× 5)

Parasola misera, Sweden (× 10)

Decaying fungi in other strange places

There is hardly any place on Earth where decaying fungi can't be found. Even at temperatures below 0 or above 50 °C there is fungal activity—and these fungi are closely studied, as they may produce enzymes with great industrial potential.

Serpula lacrymans, Denmark (SAE)

Many fungi live in our houses, but one in particular has specialized in this habitat: the dry rot fungus (*Serpula lacrymans*). This brown-spored relative of the boletes can transport water around a house through hyphal cords. It is thus able to attack otherwise dry wood, causing a dramatic brown rot. This fungus is also able to grow through brick walls, dissolving the mortar with acid. Because it has a very narrow temperature range, it has been a mystery where it could survive in nature. Recently, however, it has been discovered in mountainous regions with heavy, isolating winter snows.

An airoplane with or without *Cladosporium resinae*?

A photographic slide with a flying cattle egret and a nice fungal mycelium!

One of the more exotic habitats for fungi is the fuel tanks of aircraft. During the early sixties, microorganisms were discovered to clog fuel pumps and filters in airplanes. Researchers found a mixture of bacteria and fungi living in water droplets and decomposing the surrounding fuel. Most of the problems were caused by the mold *Cladosporium resinae* (called the kerosene fungus), which was found in up to 80 percent of the aircraft tanks. It is now controlled with fungicides.

Just as extreme as the kerosene fungus are a number of fungal species living on camera equipment. Anyone photographing in the tropics has the “chance” of a fungal mycelium growing on the lens coating inside their lenses. Likewise, old film and slides may be attacked by fungi, which form nice but frustrating mycelia on the images.

Mycorrhizal fungi

Mycorrhiza is a mutualistic symbiosis between a fungus and a plant, in which the fungal mycelium connects to the roots of the plant. The hyphae thus become an extension of the roots, helping the plant access more water and nutrients. In return, the plant delivers to the fungus large quantities of sugars derived from photosynthesis.

Mycorrhizal connections are very important to the majority of plants and to the many mycorrhizal fungi. Without mycorrhiza, plants will not thrive and fungi may be unable to grow.

Mycorrhizal roots with pink, brown, and black ectomycorrhizal fungal sheaths and white hyphal strands, Denmark (× 5)

There are two dominating types of mycorrhiza called ectomycorrhiza and A-mycorrhiza (p. 216).

Ectomycorrhiza is by far the easiest to observe. It is a connection between mostly trees (like beech, oak, birch, pine, spruce, dipterocarps, and eucalypts) and members of the Ascomycota and Basidiomycota (e.g., the genera *Amanita*, *Cortinarius*, *Helvella*, *Humaria*, *Inocybe*, *Lactarius*, *Peziza*, *Russula*, and *Tricholoma*).

In ectomycorrhiza, the fungal hyphae enter the root tips and lie between the outer cells. They also form a sheath of hyphae around the root and change the root branching pattern, making the mycorrhiza easy to observe.

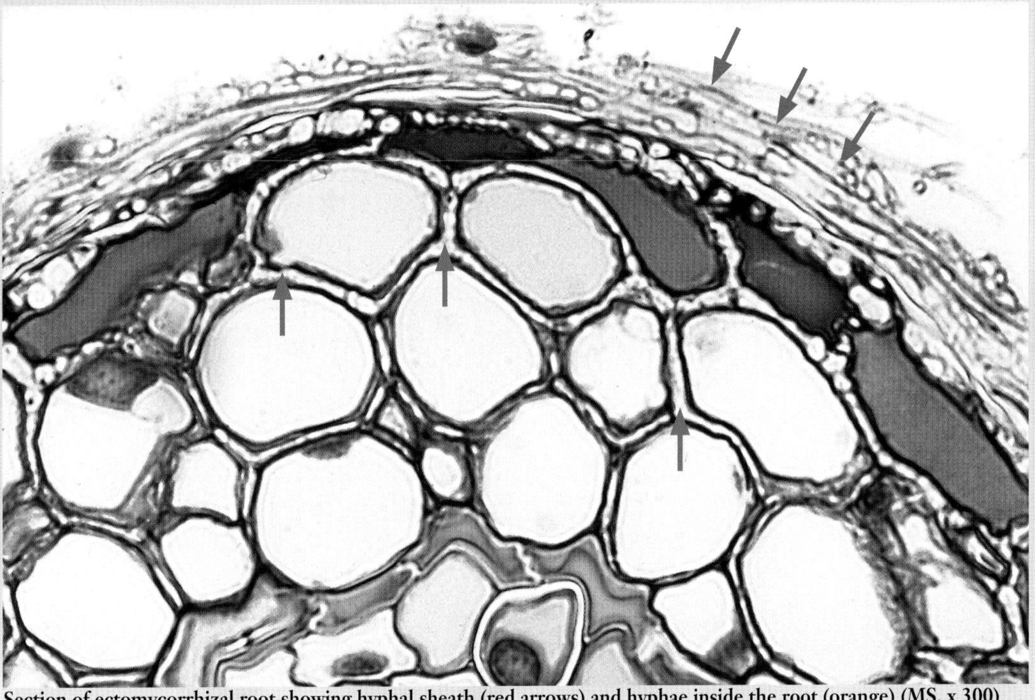

Section of ectomycorrhizal root showing hyphal sheath (red arrows) and hyphae inside the root (orange) (MS, x 300)

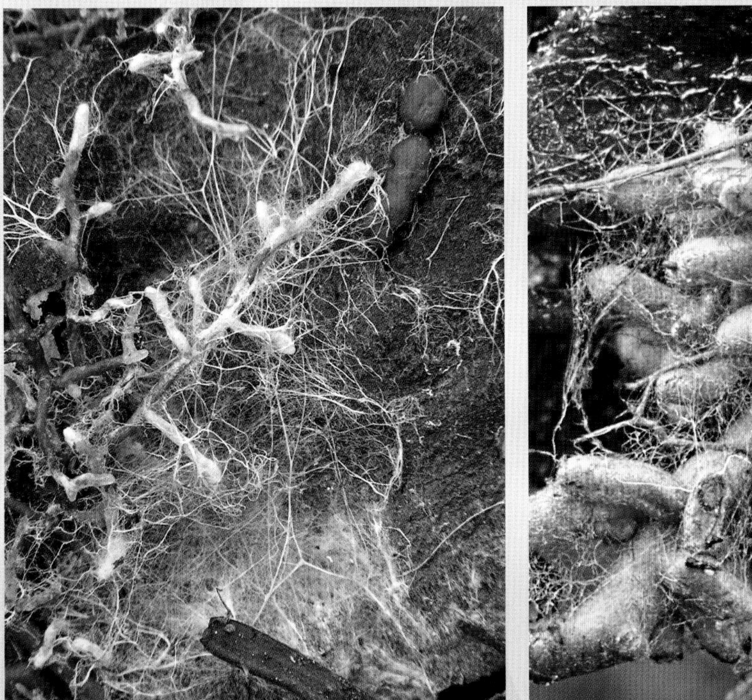

Ectomycorrhizal roots with white and yellow hyphal sheaths and hyphae radiating into the surroundings, Denmark (× 4 and 8)

Many of the available nutrients in a forest are found in dead, fallen trunks. Thus the trees surrounding decaying trunks often send roots into them and form ectomycorrhizal connections already there. When the decomposing fungi have done their job, the ectomycorrhizal fungi are in place to absorb any spilled minerals and feed their woody partners.

The ectomycorrhizal corticioid fungus *Piloderma fallax* forms yellow mycelia around roots inside a dead conifer trunk (× 4)

Many ectomycorrhizal fungi can partner with different species of trees and ectomycorrhizal trees always have several fungal partners. In this way the different species of both trees and fungi are connected into a complex network. Nutrients can flow through this network, enabling, for example, sugars from older trees to end up in shaded seedlings and minerals to cross from one fungal species to another. The network may even connect different species of trees, for example, *Picea* and *Fagus*, via the hyphae of ectomycorrhizal fungi, thus enabling compounds to cross species barriers.

The hydnoid fungus *Hydnellum aurantiacum*, Sweden

The ascomycote *Peziza badia*, Denmark

The corticioid fungus *Tomentella lilacinogrisea*, Denmark

The agaric *Amanita virosa*, Denmark

The ascomycote *Helvella crispa*, Denmark

The chanterelle *Cantharellus pallens*, Denmark

The bolete *Boletus edulis*, Denmark

The false truffle *Melanogaster broomeianus*, Denmark (IV)

A-mycorrhiza is less conspicuous but not less important than ectomycorrhiza. It forms connections between a wide range of plants on one side and relatively few fungal species belonging to the phylum Glomeromycota on the other.

A-mycorrhiza is named after the branched structures called *arbuscules* that are formed by the fungus inside the plant's root cells. By so doing, the hyphae penetrate the cell walls (the outer, structural cell layer), but not the thinner cell membranes. The penetrating hyphae then branch into a dense cloud of fungal mycelia, creating an extremely tight connection between plant and fungus. The fungus may also form globose structures called *vesicles*. To the naked eye, the mycorrhizal roots look quite normal—only special dyeing techniques will reveal A-mycorrhiza under the microscope.

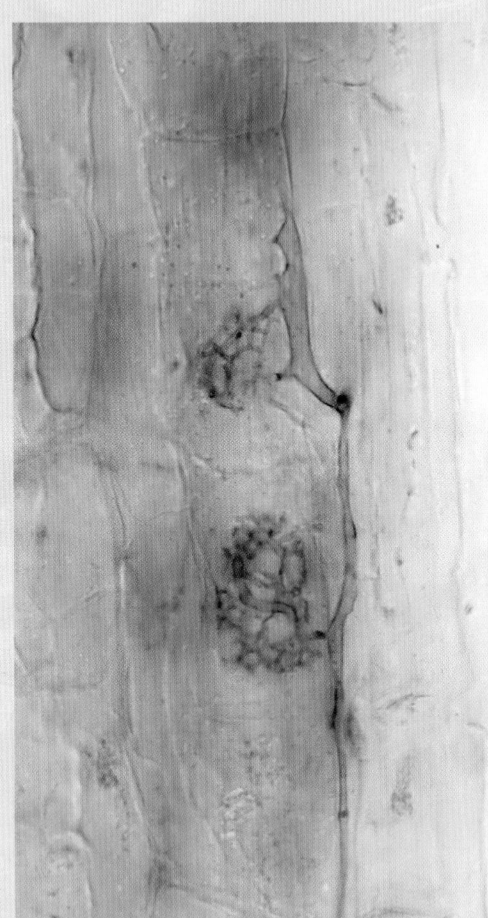

Arbuscules of A-mycorrhiza (blue) inside root (AS)

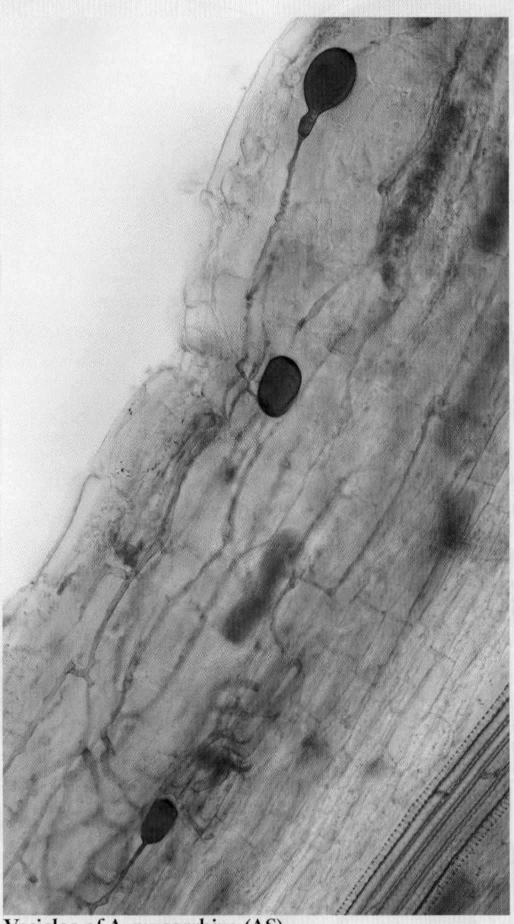

Vesicles of A-mycorrhiza (AS)

Most plant genera form A-mycorrhiza:

A-mycorrhiza is common in most major groups of plants except the cabbage family (Brassicaceae) and the goosefoot family (Chenopodiaceae). The symbiosis is especially important in providing the plants with additional phosphorus.

Most of our important crops depend on A-mycorrhiza, including cereals, most vegetables, and fruit trees, to mention a few.

The short, unbranched roots of leek need the fungal mycelium to get nutrients

The leek is one of many vegetables that form A-mycorrhiza

Cat's tail (*Phleum*)

Willowherb (*Epilobium*)

Willow (*Salix*)

Rose (*Rosa*)

Lichenized fungi

Lichens constitute another form of symbiosis between photosynthetic organisms and fungi. In this case, the photobionts are green algae or cyanobacteria, whereas the fungi involved are almost exclusively Ascomycota, mostly belonging in the class Lecanoromycetes.

The fungus runs the show. It builds the tissue (*thallus*) that houses the photobionts. This gives protection against drought and ultraviolet light and thus extends the habitat range of the photobionts. In return, the photobionts (mostly a one-celled green alga in the genus *Trebouxia*) reside in the thallus and produce sugars for the fungus by photosynthesis.

One key to the success of lichens is their ability to grow with very minimal nutrients. Growth rates of a few millimeters per year are common, as is the ability to vegetate for long periods in a completely dry and sometimes frozen state.

See more on lichens on page 88.

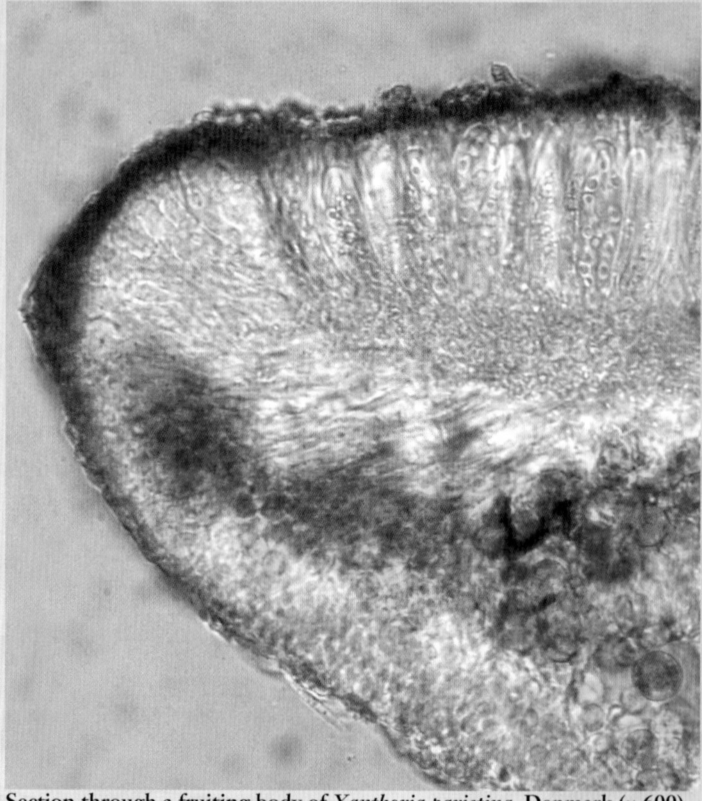

Section through a fruiting body of *Xanthoria parietina*, Denmark (× 600)

Section through the outer surface of a fruiting body of *Xanthoria parietina* showing green algae, Denmark (× 1600)

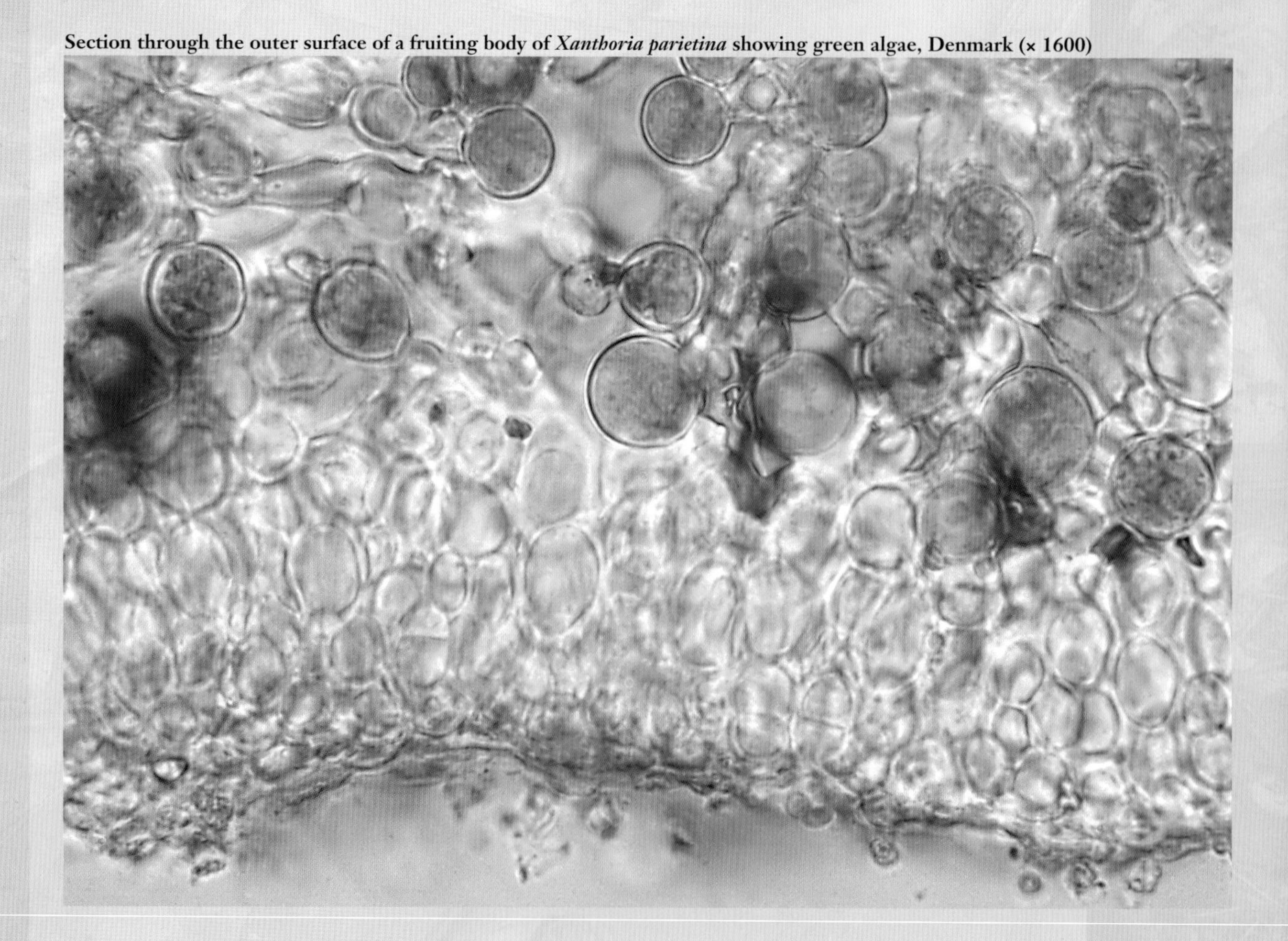

Thalli and disk-shaped fruiting bodies of *Xanthoria parietina*, Denmark (× 7)

Cladonia ramulosa—the dark tips are the fruiting bodies, Danmark (× 3)

Parasitic fungi

Parasitic fungi utilize the nutrients of a host organism without returning any "payment"—an entirely one-sided relationship.

In some cases the parasitic fungus is dependent on the host staying alive. This is the case, for example, when rusts parasitize the leaves of living plants.

In other cases the fungus kills the host and absorbs the remnants. This is the case when honey fungus (*Armillaria*) or root rot (*Heterobasidion*) kills a tree, or when a species of *Cordyceps* devours an insect (see page 81).

The first type of parasite typically does much harm to our agricultural crops, while the second is a problem in forestry. However, parasites can also be helpful in controlling other problematic organisms (weeds, insects, and fungi) in agriculture.

Puccinia punctiformis—a parasite on *Cirsium*, Denmark

Heterobasidion annosum—a parasite on conifers, Denmark

Armillaria mellea—a parasite on broad-leaved trees, Denmark

Fungi in the world

Though they live a hidden life, fungi influence the world in numerous ways. The *funga* (the fungal equivalent of an area's *flora* and *fauna*) is important in ecosystems everywhere. From the cold shores of the Antarctic to tropical rainforests, from human habitations to pristine virgin forests—fungi are there!

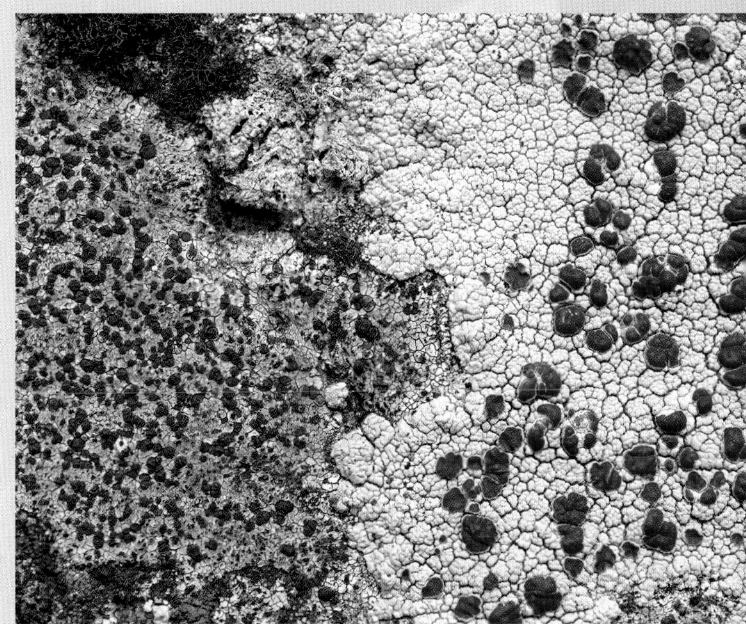

Lichens on a subarctic stone, Sweden (× 2)

Where fungi grow

The funga under the harsh conditions of *the High Arctic* may be dominated by lichens, but during the short summer the more fleshy fruiting bodies of agarics and cup fungi can also be numerous. In subarctic areas, a diverse ectomycorrhizal funga is present in symbiosis with, for example, shrubs of *Betula*, *Salix*, and *Dryas*, and the herb *Polygonum viviparum*.

Myriosclerotinia ciborium parasitizing *Eriophorum angustifolium*, Greenland (× 4)

Amanita groenlandica towering over the mycorrhizal partner dwarf birch (*Betula nana*), Greenland

Where fungi grow . . .

Boreal coniferous forests cover the northern portions of North America, Europe, and Asia just south of the subarctic zone. This area may be alive with fungi before the short, hectic autumn ends in an early frost.

Though there are numerous decomposing fungi, the ectomycorrhizal fungi that grow with, for example, *Picea*, *Pinus*, *Larix*, and *Tsuga* often dominate the funga. Genera like *Cortinarius*, *Russula*, *Lactarius*, *Inocybe*, *Hebeloma*, *Sarcodon*, and *Ramaria* are abundant here.

Suillus asiaticus growing with larch (*Larix*), Russia (TL)

Sarcodon imbricatus in mixed forest with spruce (*Picea*) and pine (*Pinus*), Sweden (JV)

Cortinarius caninus growing with spruce (*Picea*), Sweden

Where fungi grow . . .

Temperate deciduous forests lie south of the coniferous zone in the Northern Hemisphere. Again, decomposing and ectomycorrhizal fungi are present in great numbers. The growing season here is normally long, and in wet, mild years it is virtually year-round.

These forests are dominated by ectomycorrhizal trees, so the mycorrhizal species are a very important component of the funga. On the other hand, the annual leaf fall produces a vast amount of biomass to be recycled every year, so litter decomposers are also present in great numbers.

Laccaria amethystina forming ectomycorrhiza with beech (*Fagus*)

Mycena crocata decomposing beech (*Fagus*) wood

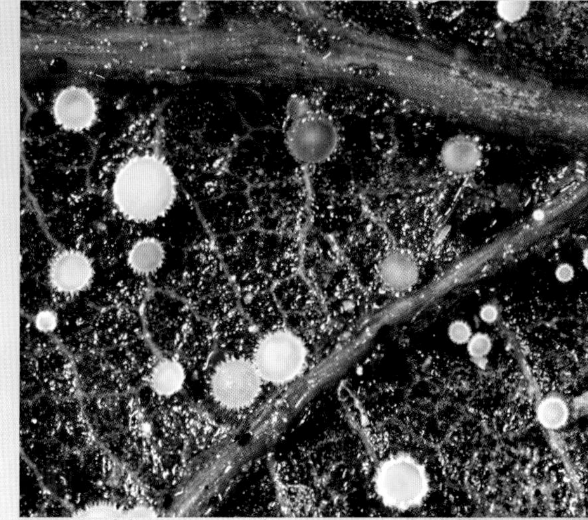

Mosaik of discomycetes decomposing an oak leave, (× 10)

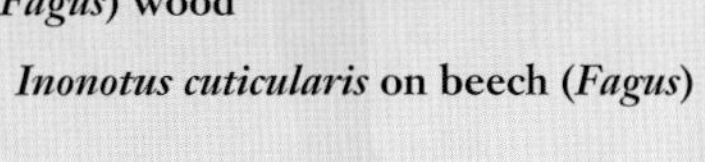

Inonotus cuticularis on beech (*Fagus*)

Where fungi grow . . .

Dry evergreen scrubland is found in many subtropical areas as well as in the drier tropics. The fungal season generally follows a rainy period; in Southern Europe it occurs during late autumn and winter.

While still hosting many ectomycorrhizal species, this funga also includes a high number of drought-tolerant fungi. These may include genera like *Daldinia* and *Hypoxylon*, or even small cup fungi like those in the genus *Orbilia*.

Orbilia coccinella (× 10)

Daldinia vernicosa (× 1)

Hymenochaete sp., Ecuador (× 2)

Where fungi grow . . .

Steppe and desert—who would think fungi could grow there? But they do! Species of *Termitomyces* fruit on termite nests, agarics are reduced to puffball-like structures, dusting their spores away, and ectomycorrhizal truffles lie protected in the soil. And after the rains, a funga of numerous small agarics may appear in the short window before everything dries up again.

Termitomyces fuliginosus, Burkina Faso (CL)

The *Coprinus*-like gastromycete *Podaxis pistillaris*, Burkina Faso

Pisolithus sp. growing with eucalyptus, Burkina Faso

Digging for desert truffles (*Terfezia* sp.), Egypt (JF)

Where fungi grow . . .

Wet mountain forests and *cloud forests* are characterized by very long moist periods when the vegetation is more or less constantly immersed in clouds. This yields an extremely lush funga, often with numerous species of both mycorrhizal and decomposing fungi. Also, trees are covered with epiphytic plants, mosses, and lichens.

Depending on the local climate, the fruiting season may be year-round or it may be connected to seasonal rains, as in the Himalayas, where the monsoon starts around June and the fungal season may stretch into November.

Clavaria sp., Ecuador (× 4)

Amanita rubrovolvata, Bhutan

Dacrymyces sp., Bhutan (× 4)

Hygrocybe sp., Ecuador (× 2)

Where fungi grow . . .

Tropical rainforests are the peak of all biodiversity, including an extremely diverse funga. While the more or less constantly wet forest ensures the continuous presence of fruiting bodies, the lack of pronounced seasons may suppress the impressive fruiting peaks found in more seasonal climates.

The Amazonian rainforest may be rather lacking in ectomycorrhizal fungi; these may be more common in Africa and Asia. Genera like *Gymnopus*, *Cookeina*, *Favolaschia*, *Marasmius*, *Mycena*, and *Xylaria* are among the dominant decomposers, seconded in ectotrophic regions by russulas, amanitas, chaterelles, and others.

Marasmius ruforotula, Ecuador (× 2)

Marasmius rhyssophyllus, Ecuador

Xylaria hyperythra, Ecuador

Favolaschia sp., Ecuador (× 15)

Cookeina speciosa, Ecuador

Where nothing grows

There is one habitat in the world where the fungi have no chance: the agricultural desert. Here fertilizers and fungicides put an end to almost all kinds of fungi, inclucing the mycorrhizal species that live in the soil and that would normally

help the plants stay in good health and produce larger crops. One dose of fertilizer can destroy a grassland habitat for sensitive species of *Hygrocybe*, *Entoloma*, and *Geoglossum* for many decades—constant spraying with fungicides will take the rest . . .

Why we need fungi . . .

Fungi are beautiful. This was already realized two thousand years ago when the Roman town Herculaneum was covered by the erupting Vesuvius volcano—and it is known today by nature lovers, artists, and mycophiles, some of whom go to extremes, tattooing fungal shapes on their skin.

The first rule in nature conservation should be: We need nature because it inspires us and we love it . . .

The 2000-year old Herculaneum fresco of *Lactarius deliciosus*

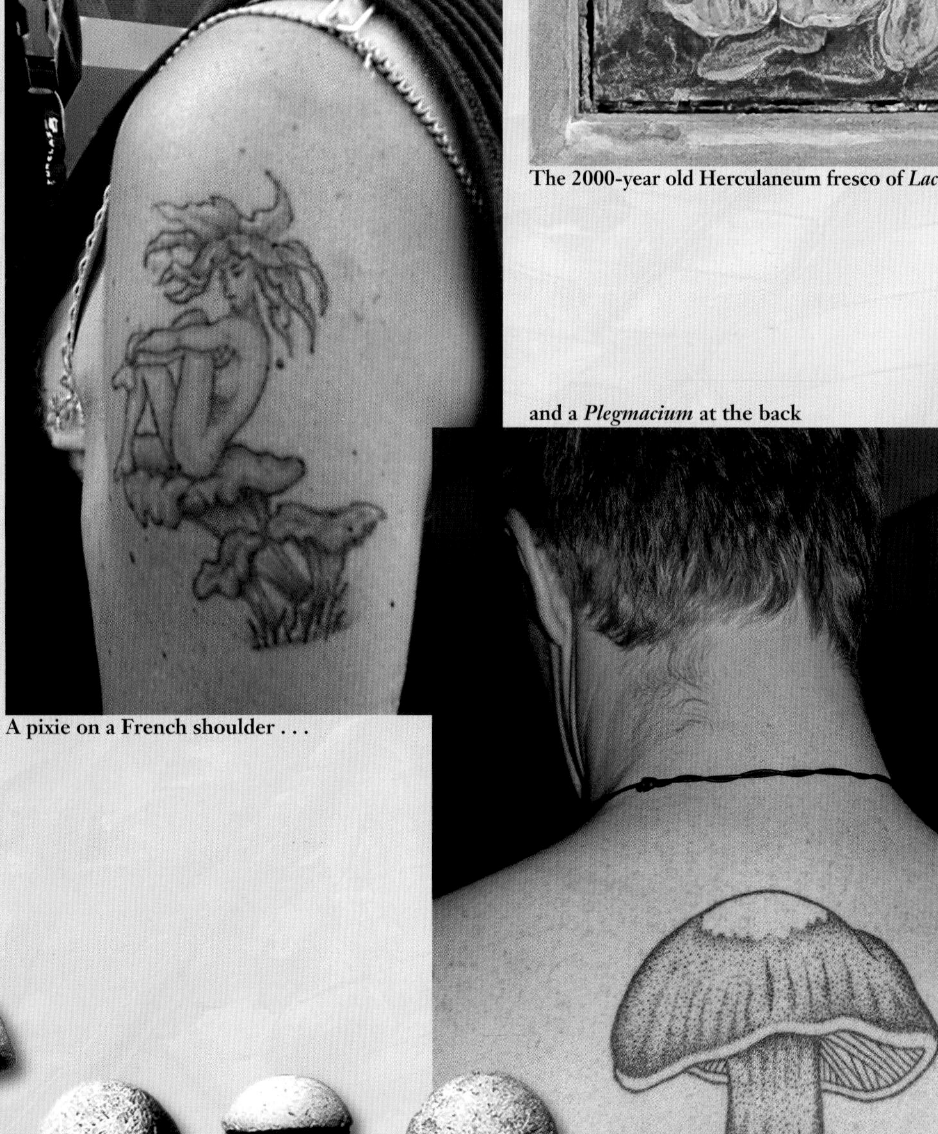

A pixie on a French shoulder . . .

and a *Plegmacium* at the back

Ancient South American mushroom statues

Beautiful, elegant, mysteriously short-lived fruiting bodies of *Coprinopsis macrocephala*, Denmark (× 8)

Why we need fungi . . .

Fungi are fascinating. **Mycophagists and mycologists—those who eat and those who study fungi—may be caught performing with these fascinating creations . . .**

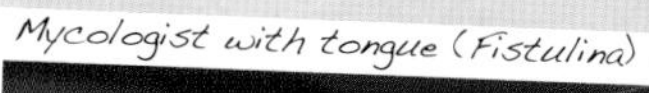
Mycologist with tongue (Fistulina) . . .

These are mine! (MC)

The largest bolete . . .

Can this be smoked?

Student with Fomes antlers

Mycologist juggling Podaxis

Collecting towards Mecca

The largest agaric of the world? (RA)

Does it really smell like hyacinths?

Why we need fungi . . .

Fungi remove dead organic material. If we didn't have fungi, plant remains, especially wood, would slowly pile up in nature. The very elaborate enzyme systems of fungi are superior in breaking down the complex cell walls of plants.

Pholiota adiposa and others, Denmark

Why we need fungi . . .

Fungi stabilize and serve ecosystems through mycorrhizal connections. Without the fungi to work as extended root systems, many plants would suffer from malnutrition and our conventional agriculture would demand much higher levels of fertilizers, leading to greater CO_2 emissions and even more pollution.

In addition, mycorrhizal fungi connect different plant species and different fungi

into a complex network. Through these mycorrhizal connections, nutrients may be exchanged across species boundaries, resulting in a stabilized ecosystem with high biodiversity.

Roots colonized by different ectomycorrhizal fungi (× 10)

Why we need fungi . . .

Fungi serve as an important food source in many societies. **As fungi are more related to animals than to plants, they contain some nutrients that are otherwise hard to get in a vegetarian diet. Likewise, fungi may contain vitamins and antioxidants. Among the most esteemed wild fungi are species of *Amanita*, *Boletus*, *Cantharellus*, *Craterellus*, *Lactarius*, *Morchella*, *Termitomyces*, and *Tricholoma*.**

Some fungi are eaten for their psychoactive properties. In particular, some species of the genus *Psilocybe*—known as magic mushrooms, or just shrooms—are valued for their LSD-like effects. These fungi have been used around the world in religious ceremonies for thousands of years. However, in many modern societies, possession of this particular piece of nature is deemed illegal.

The valuable *Tricholoma matsutake*, Thimphu market, Bhutan

Negotiating for chicken of the woods (*Laetiporus* sp.), Bhutan

Mushroom hunters, Denmark

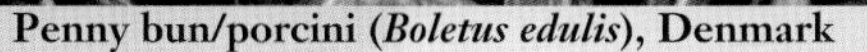

Penny bun/porcini (*Boletus edulis*), Denmark

Tuber aestivum, Denmark (JV)

The liberty cap (*Psilocybe semilanceata*), Denmark

Some edible fungi are grown commercially. Especially in Europe, North America, and Asia, species of *Agaricus*, *Auricularia*, *Pleurotus*, *Lentinula*, *Flammulina*, and *Volvariella* are grown in large quantities. Unfortunately, the most delicious fungi are ectomycorrhizal and don't produce fruiting bodies in culture.

Pleurotus ostreatus cultivation, Denmark

Selling cultivated *Volvariella volvacea*, Thailand (FR)

Wild mushrooms are picked and eaten around the world. In Africa and Asia, fungi are collected and sold in markets and on roadsides. These fungi may be an important part of the local economy and food base. In eastern and southern Europe there is also a long tradition of collecting wild edible fungi, while the tradition in western Europe and the Americas is for the most part more recent.

The American mycologist and photographer David Arora has traveled the world studying ethnomycology. Here is a peek at his findings . . .

Collecting wild morels (*Morchella*), USA (DA)

Ahmed Soumitte with porcini (*Boletus edulis*), USA (DA)

A gold digger: Cowboy Randex with chantarelles, USA (DA)

Collecting edible amanitas in Mexico (DA)

Mayan Indian sorting the days catch, Mexico (DA)

Wild white truffles (*Tuber magnatum*) on display in Italy (DA)

Yana, Anya, Vova, and Vadik with an *Amanita*, Russia (DA)

Yunnan girl with *Boletus edulis* and flowers (DA)

Termitomyces sp. offered at a market in Asia (DA)

Picking *Volvariella volvacea* in the Maluku Islands (DA)

Roadside sales of wild fungi, Malawi (DA)

A catch of *Termitomyces titanicus*, Botswana (DA)

Why we need fungi . . .

Fungi have been praised for their medicinal qualities for many centuries. During the Middle Ages, lichens growing on human skulls—under the name "muscus cranii humani"—were used to cure epilepsy. Medicine has evolved since, but the use of many fungi is still heavily debated.

In Asian medicine a number of fungal species like the reishi mushroom (*Ganoderma lucidum* agg.), shiitake (*Lentinula edodes*), and yartsa gunbu (*Cordyceps sinensis*) are used to cure a wide range of diseases from depression and immune system deficiencies to cancer. Western medicine holds a rather skeptical view, but much research is being done in the field.

Reishi mushrooms (*Ganoderma lucidum* agg.) (MG)

Shiitake (*Lentinula edodes*) cultivated on oak, Denmark

Moth larvae with *Cordyceps sinensis* (NHJ)

Penicillium kill bacteria around their mycelia (FG)

Less controversial is the use of various fungi for the production of antibiotics. The original discovery of modern antibiotics was the penicillin produced by a species of the ascomycote genus *Penicillium*. The effect of *Penicillium* on bacteria in a petri dish was discovered in 1928, but it was not until World War II that a drug against infections became available.

"Muscus cranii humani"—lichens on a human skull, Greenland (US)

Catathelasma sp. sold for medicine in Bhutan

Why we need fungi . . .

Fungi serve as sources and factories for the production of enzymes and medicines. Following the success of antibiotics, fungi have become a major source of other valuable chemical compounds. The ability of fungi to decompose organic compounds is based on specific enzymes. The use of suitable fungal enzymes for washing in cold water, for example, may save substantial amounts of energy. Also, the idea of using organic waste products for biofuel is based on fungal enymes.

Peniophora lycii, Denmark (IV)

To find the relevant enzymes, thousands of fungi are collected, cultured, and screened for enzymatic activity. This may lead to strange discoveries, such as that of the corticiaceous wood-decaying fungus *Peniophora lycii* producing an enzyme able to liberate phosphorus in plant material. Traditionally the feed of some animals, including pigs, was enriched with phosphorus, resulting in environmental pollution. Mixing the phytase produced by *Peniophora* into pigs' feed enables them to utilize the phosphorus already present in the plants, averting the necessity of adding phosphorus and the resulting leakage.

For greater efficiency, other fungi can be utilized as "biofactories" to produce the desired compounds. In the case of phytase, the gene coding for this enzyme was engineered into the production fungus *Aspergillus oryzae*. This fungus is then cultivated in large quantities, producing the phytase as a by-product during its growth.

Pseudoplectania nigrella, Denmark

Strobilurus esculentus, Denmark

The search for biochemically interesting fungi goes on. What started with a search for metabolites in asexual fungi like *Penicillium* has now spread to fungi with larger fruiting bodies. Interesting antibiotics have recently been found in the cup fungus *Pseudoplectania nigrella* and the agaric genus *Strobilurus*.

Fungal future

Considering the short and sporadic occurrence of many fungal fruiting bodies and the rather small amount of research conducted so far in mycology compared to zoology and botany, the present knowledge of fungal diversity is obviously incomplete. For example, it is quite easy to obtain soil samples containing unknown fungal DNA from even the most well-known locales.

To place fungi on the biological world map we need a realistic estimate of the number of fungal species in nature. In the UK, there are close to 8,000 known species of fungi (this number increases by about 40 per year) and around 1,450 species of higher plants. In other words, the number of known fungi is close to six times the number of higher plants. This ratio must be an absolute minimum, as the numerous expected but yet undiscovered species are not included.

Ecuador is a relatively well-researched tropical country. At present we know at least 16,000 species of higher plants from Ecuador. Consequently, a prediction based on the numbers in the UK will give around 100,000 species of fungi. The present number of known fungal species is 3,800, or 3.8 percent of the expected species number.

For planet Earth, recent estimates range from 1.5 to 5 million fungal species. At present only about 100,000 fungal species are described. Taking the conservative point of view, this leaves:

**1.4 million species,
or 93 percent, still to be discovered!**

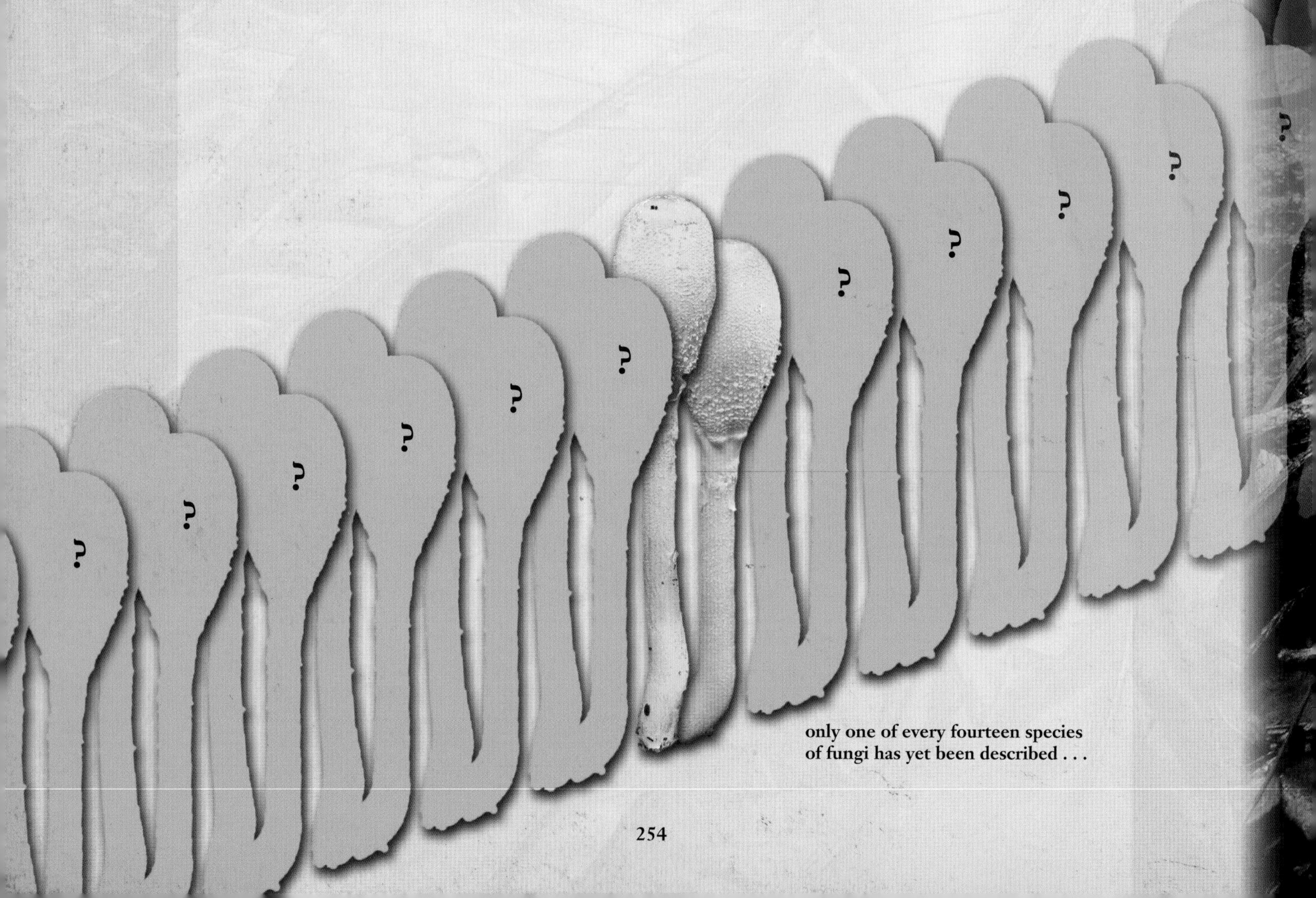

only one of every fourteen species of fungi has yet been described . . .

Protecting the fungi

To ensure the availability of future fungal biodiversity we should:

1. Work against rapid climate change.
2. Protect the biodiversity and continuity of forests.
3. Manage grazed and unfertilized grassland.
4. Stop the uninhibited use of fertilizers and fungicides in agriculture and forestry.
5. Encourage research in fungal taxonomy and biology.

Morning over the Amazon rainforest—one of the world's great pools of undiscovered biodiversity . . .

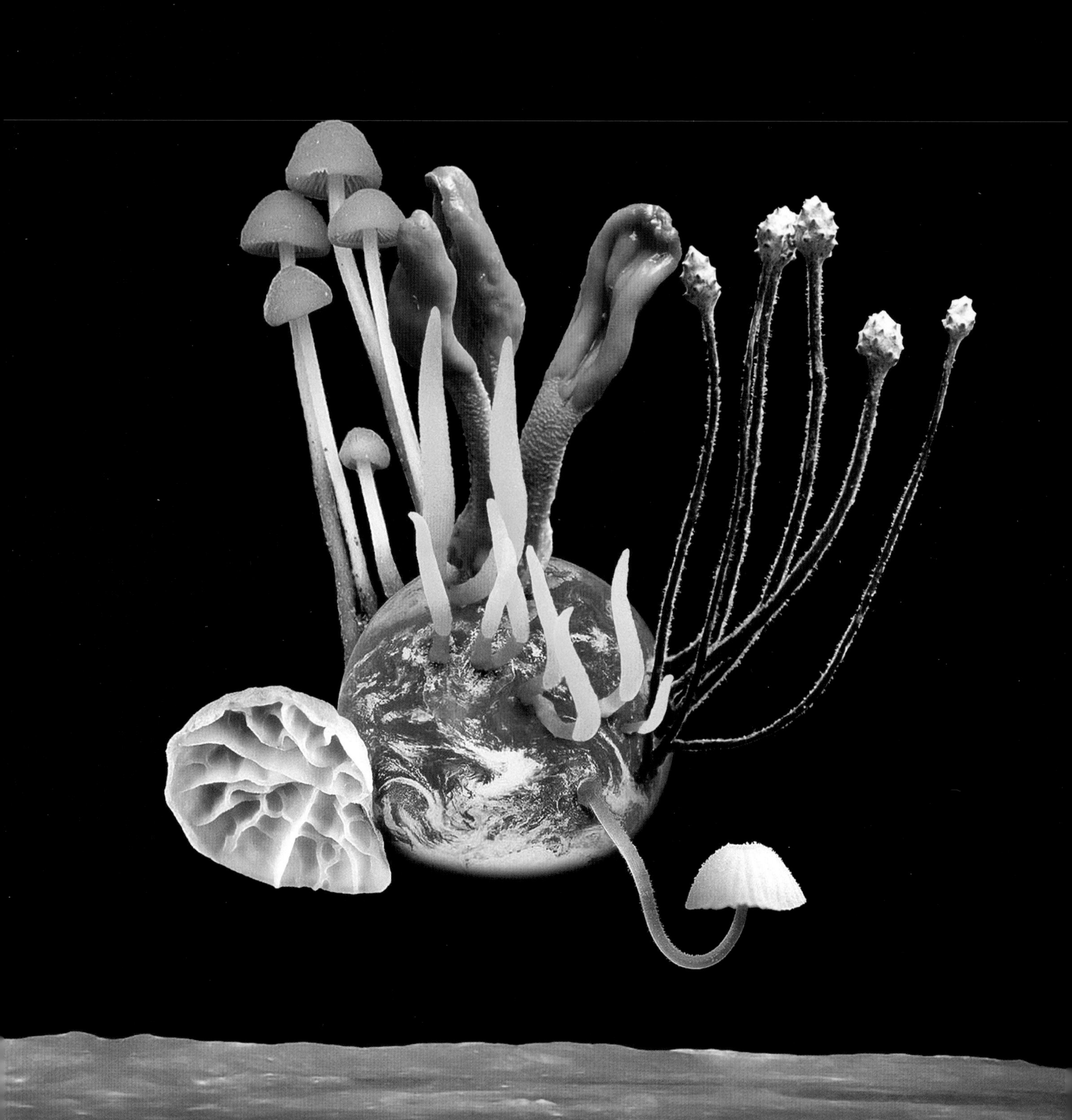

Postscript

The "Amazonian Mystery Tongue" was found on several occasions during field-work in Ecuador. This fungus was believed to belong to the very conspicuous genus *Guepinia*, which belongs in the jelly fungi. Later examination, however, revealed that its basidia had transverse walls, thus placing it outside any known genus.

This example illustrates our knowledge of fungal diversity. On a single week-long visit to the Amazonian rainforest, several collections of striking but seemingly undescribed species can be made. Even in small, well-known countries in Europe—where fungi have been studied for several centuries—undescribed species are routinely discovered.

The "Amazonian Mystery Tongue" might be the fungus with the enzyme system that could produce a cheap and efficient transformation of straw into biofuel or an agent against cancer. With today's rate of species extinction, however, it may go extinct before we even get to know it . . .

"The Amazonian Mystery Tongue", Ecuador 2002

Index

Marasmius sp., Ecuador

The Kingdom of Fungi

Published by Princeton University Press, 41 William Street, Princeton, New Jersey 08540
In the United Kingdom: Princeton University Press, 6 Oxford Street, Woodstock, Oxfordshire OX20 1TW
nathist.princeton.edu

Published in Denmark by Gyldendal A/S, Klareboderne 3, 1001 København K, Denmark
www.gyldendal.dk

Library of Congress Cataloging-in-Publication Data

Petersen, Jens H., 1956-
The kingdom of fungi / Jens H. Petersen.
p. cm.
Includes bibliographical references and index.
ISBN 978-0-691-15754-2 (hardcover : alk. paper) 1. Fungi. I. Title.
QK603.P48 2013
561′.92—dc23 2012031873

British Library Cataloging-in-Publication Data is available

This book has been composed in Janson Text

Design and composition by Jens H. Petersen

Printed on acid-free paper. ∞

Printed in Slovenia

10 9 8 7 6 5 4 3 2 1

Illustrators

All illustrations are by the author, except:
René Andrade (RA): 5, 241
David Arora (DA): 191, all photos on 246–247
Pia Boisen Hansen (PBH): 87
Carsten Brandt (CB): 5
Morten Christensen (MC): 240
Steen A. Elborne (SAE): 208
John Feeney/ Saudi Aramco World/SAWDIA (JF): 231
Jens Frisvad and Susanne Gravesen (FG): 251
Tobias Frøslev (TF): 64
Marija Gregori (MG): 250
Hans Hillewaert/Wikipedia (HH): 22
Anette Højlund (AH): 9
Jimmie Høier (JH): 101
Nigel Hywel-Jones (NHJ): 250
Thomas Læssøe (TL): 14, 77, 106, 120, 121, 144, 168, 176, 225
Christian Lange: 231
Flemming Rune (FR): 247
Martin St-Amant (MSA): 201
Mikako Sasa (MS): 211
Anni Sloth (AS): 102, 216
Ulrik Søchting (US): 251
P. van der Valk and J.G.H. Wessel (VW): 195
Jan Vesterholt (JV): 58, 64, 98, 150, 214, 225, 247, 252
Trude Vrålstad (TV): 42

Thanks to Thomas Læssøe for his friendship, company, and mycological sparring on numerous field trips and especially for reading and commenting on this manuscript. Also thanks to the photographers who lent me their unique pictures and to all you mycologists out there who keep the pot boiling.

Learn more about fungi on www.mycokey.com